SpringerBriefs in Applied Sciences and Technology

Computational Mechanics

Series Editors

Holm Altenbach⊙, Faculty of Mechanical Engineering,
Otto-von-Guericke-Universität Magdeburg, Magdeburg, Sachsen-Anhalt, Germany

Lucas F. M. da Silva, Department of Mechanical Engineering, Faculty of
Engineering, University of Porto, Porto, Portugal

Andreas Öchsner, Faculty of Mechanical Engineering, Esslingen University of
Applied Sciences, Esslingen, Germany

These SpringerBriefs publish concise summaries of cutting-edge research and practical applications on any subject of computational fluid dynamics, computational solid and structural mechanics, as well as multiphysics.

SpringerBriefs in Computational Mechanics are devoted to the publication of fundamentals and applications within the different classical engineering disciplines as well as in interdisciplinary fields that recently emerged between these areas.

More information about this subseries at https://link.springer.com/bookseries/8886

Alireza Akhavan-Safar · Eduardo A. S. Marques ·
Ricardo J. C. Carbas · Lucas F. M. da Silva

Cohesive Zone Modelling for Fatigue Life Analysis of Adhesive Joints

 Springer

Alireza Akhavan-Safar
Institute of Science and Innovation
in Mechanical and Industrial Engineering
(INEGI)
Porto, Portugal

Eduardo A. S. Marques
Institute of Science and Innovation
in Mechanical and Industrial Engineering
(INEGI)
Porto, Portugal

Ricardo J. C. Carbas
Institute of Science and Innovation
in Mechanical and Industrial Engineering
(INEGI)
Porto, Portugal

Lucas F. M. da Silva
Department of Mechanical Engineering
Faculty of Engineering
University of Porto
Porto, Portugal

ISSN 2191-530X   ISSN 2191-5318 (electronic)
SpringerBriefs in Applied Sciences and Technology
ISSN 2191-5342   ISSN 2191-5350 (electronic)
SpringerBriefs in Computational Mechanics
ISBN 978-3-030-93141-4   ISBN 978-3-030-93142-1 (eBook)
https://doi.org/10.1007/978-3-030-93142-1

This Springer imprint is published by the registered company Springer Nature Switzerland AG
The registered company address is: Gewerbestrasse 11, 6330 Cham, Switzerland

Contents

1 Introduction 1
    1.1 Adhesive Joints 1
    1.2 Fatigue of Bonded Structures 2
    1.3 Fatigue Initiation Versus Fatigue Crack Propagation 3
    1.4 High Cycle Fatigue (HCF) Versus Low Cycle Fatigue (LCF) 4
    1.5 Fatigue Characterization 5
    1.6 Factors Affecting the Fatigue of Adhesives 6
        1.6.1 Joint Geometry 6
        1.6.2 Effect of Material Property 7
        1.6.3 Effect of Loading Conditions 7
        1.6.4 Environmental Effects 8
    1.7 Fatigue Threshold Energy 10
    1.8 Force Versus Displacement Control Tests 10
    1.9 Life Prediction Approaches 11
        1.9.1 Total Fatigue Life (S–N) Approaches 11
        1.9.2 Fatigue Crack Growth Method 11
        1.9.3 Damage Models 13
    1.10 Numerical Methods 14
        1.10.1 Extended Finite Element Method, XFEM 14
        1.10.2 Interfacial Thick Level Set Modeling 14
        1.10.3 CZM 15
    References 15

2 Cohesive Zone Modelling-CZM 19
    2.1 History and Concepts 19
    2.2 Traction Separation Laws 21
        2.2.1 Bilinear TSL 22
        2.2.2 Trapezoidal TSL 24
        2.2.3 Linear-Exponential TSL 24
        2.2.4 Multi-linear Softening TSL 24
        2.2.5 Customized TSL 25

        2.2.6   Which TSL Should Be Used? 26
   2.3   How to Identify the CZM Parameters? 28
        2.3.1   Classic Characterization Method 28
        2.3.2   Inverse Method 29
        2.3.3   Direct Method 29
   2.4   CZM Constitutive Laws 30
        2.4.1   Pure Mode Conditions 30
        2.4.2   Mixed Mode Constitutive Laws 35
   2.5   Fatigue Analysis Using CZM 39
   References 40

3 Fatigue Degradation Models 43
   3.1   Introduction 43
   3.2   Loading Envelope Strategy 45
        3.2.1   Linking Damage and Fracture Mechanics Approach
               (LDFA) 45
        3.2.2   Experimental Data Fitting (EXFIT) 52
   3.3   Mixed Mode Paris Laws 56
   3.4   Variable Amplitude Fatigue Loading 58
   3.5   Cycle-By-Cycle or Unloading–Reloading CZM 60
   References 63

4 Numerical Simulation 67
   4.1   Introduction 67
   4.2   Commercial Software 68
   4.3   CZM 68
        4.3.1   Parameters Adjustments 69
        4.3.2   Cohesive Contact Versus Cohesive Element 72
        4.3.3   Mesh Size Sensitivity 74
   4.4   Fatigue Analysis 74
        4.4.1   Cycle by Cycle Versus Cycle Jumping 74
        4.4.2   User Subroutines 76
   4.5   Examples of Numerical Models 80
   References 87

5 Summary and Conclusions 89
   5.1   Summary 89
   5.2   Conclusions 92
   References 93

Chapter 1
Introduction

Abstract Considering the fatigue response of adhesive joints is an essential part of a joint design since, based on industrial reports, 90% of failures in industrial settings are caused by fatigue damage. The total fatigue life is divided into two main stages, naming the initiation of fatigue life and the fatigue crack propagation. Although as a conservative approach in most applications the crack initiation life is considered as the total fatigue life, for lighter structures and to reduce the costs, the crack growth fatigue life of adhesive joints should be also considered in joint design. The process of fatigue in bonded structures can be analyzed experimentally, numerically or in a combination of both. Experimental methods can be divided into two groups, including fatigue characterization using standard or routine testing and fatigue life analysis of real components. Several factors also affect the joint response, such as loading parameters, joint geometry, and environmental conditions. All of these aspects are discussed in this chapter of the book.

1.1 Adhesive Joints

Synthetic adhesives were introduced around 1900. Since then, the application of adhesives to bond structural and non-structural components has vastly increased. Today, adhesives are used to bond a wide variety of materials and components. The growth in use and popularity to which adhesives have recently been subjected is not unfounded. Compared to traditional bonding techniques, adhesives offer better damping properties due to their polymeric nature and can also be used to join very thin and dissimilar adherends. Although the strength of adhesives is significantly lower than that of metals, in the case of thin metal plates, the bond strength they allow for is greater than that of adherends, which means that they are often not the weak link in the joint. The flexibility of the joint design increases with bonding and, for example, adhesive bonding is one of the few techniques that can also provide a smooth surface to meet aerodynamic requirements in aeronautical applications, as is the high strength-to-weight ratio. All these are crucial factors that make adhesives suitable

1
A. Akhavan-Safar et al., *Cohesive Zone Modelling for Fatigue Life Analysis of Adhesive Joints*, SpringerBriefs in Computational Mechanics, https://doi.org/10.1007/978-3-030-93142-1_1

for use in light structures, but adhesives can also be found in diverse applications and in many different industrial sectors.

Due to the excellent fatigue resistance of adhesives, authors have extensively analyzed the fatigue response of bonded structures. Fatigue failure is considered the main reason for most structural failures in engineering applications and, consequently, considering fatigue damage is therefore a critical part of a joint design, especially when the bonded joint is used as a load-bearing structure. Although methods for the prediction of the static strength of bonded joints are well developed, the same cannot be said for the fatigue failure of bonded structures. In fact, understanding the fatigue failure mechanism of adhesive joints is always a challenge, due to the complexity of joint geometries, the presence of different materials with different properties in a bonded joint and the many parameters that usually affect the results.

Despite the extensive work done on this topic, fatigue failure analysis of bonded joints is still quite difficult and a reliable universal fatigue life prediction in real-world applications remains an open topic of research. Since quality control and assessment techniques for bonded joints are not as well developed as those available for metals and even composites, fatigue failure analysis of such structures has become crucial, especially for joints which have been designed based on the concepts of safe life or of damage tolerance.

1.2 Fatigue of Bonded Structures

As discussed above, adhesive joints, provide better fatigue life when compared with bonded joints. However, the fatigue of bonded structures is one of the most common types of failure that bonded areas can experience in service. Fatigue failure, which is related to specimens subjected to cyclic loading, generally occurs at load levels much lower than those corresponding to the static strength of the joints. Based on diverse reports, over 90% of component service failures in industries are related to fatigue damage [1] and even in advanced and critical structures where the joints must pass several quality control requirements, fatigue failure of bonded joints is ultimately unavoidable. For example, on F-111, Airbus A310, and CF-5 aircraft, fatigue cracking in bonded structures has been observed during service [2, 3], a clear signal that fatigue failure mechanisms of adhesive joints are still far from fully understood.

Several parameters can significantly influence the fatigue response of adhesives such as the joint geometry, the interface condition, in-service loading, and environmental parameters. Accordingly, for a reliable design, especially for primary load-bearing structures, a comprehensive understanding of fatigue failure and life prediction mechanisms is indispensable, which can often only be attained with an extensive experimental program. However, experimental characterization of the fatigue damage of adhesives is an often costly and time-consuming endeavor. An alternative is to use a numerical approach combined with a few experimental results to estimate the fatigue life of bonded joints. Numerical fatigue analysis using life prediction models is cost-effective and allows engineers to reduce the number and types of tests needed to estimate the fatigue life of real joints. Furthermore, numerical parametric studies can also be employed to optimize the joint geometry and materials, always targeting enhanced fatigue performance.

1.3 Fatigue Initiation Versus Fatigue Crack Propagation

The total fatigue life of bonded joints is divided into two different stages. The first is the onset of fatigue cracks and the second is the fatigue crack propagation process. Despite the different definitions presented by the authors, there are still some uncertainties about the portion of the fatigue life of adhesive joints that, in practice, takes place in the initiation phase. A parameter that is extensively used to distinguish the two stages of the fatigue life is known as the threshold energy. The life that is spent before reaching the threshold energy is considered as the fatigue onset life and the remainder corresponds to the fatigue crack propagation. Some authors believe that the fatigue crack initiation phase and propagation life can be separated by fatigue damage size. Below a specific damage size, the life can be considered as the initiation life and above this specific damage value, life is considered as the crack propagation life [4]. Industries often define the initiation phase as the lifetime of the joint spent before detecting a failure (for example microcracks). However, this definition also presents some uncertainties, as the damage detection approach and its accuracy can significantly influence the results. The presence of early flaws within the thin layer of the adhesive layer (which is often not visible and inaccessible) creates additional difficulties in analyzing the onset of fatigue of adhesive joints. Considering these points, accurately determining the onset of fatigue life in adhesive joints is still a challenge and not well developed, thus requiring further research. Several techniques have been proposed to support this determination process, such as the back face strain gauge technique, using video damage detection methods, and acoustic techniques.

In some studies, it was shown that the total fatigue life of the tested SLJs is spent mainly in the crack propagation stage, while other research works postulate that the total fatigue life is actually mainly governed by the initial life [5, 6]. Some authors also showed that reducing the fatigue load level by around 50% of the maximum static strength of the joints can lead to an infinite life, a condition where the total fatigue life is almost fully spent in the fatigue initiation phase [7].

It is important to be aware that not only the material, but also the loading conditions and joint geometry influence the initiation and propagation of fatigue damage. Adherent treatment is also another parameter that can alter fatigue initiation and propagation lives, as shown by Lefebvre and Dillard [8]. When considering aluminum epoxy wedge specimens, they found that substrate surface treatment influences the fatigue onset phase more than the propagation life.

Accordingly, to simplify the analysis, and based on the concepts of fracture mechanics, the initiation phase is often ignored by many authors and only the fatigue crack propagation life of the bonded joints is analyzed. This approach can lead to an underestimated fatigue life prediction, especially if the fatigue life is spent mainly in the initiation stage, e.g. for joints with more brittle adhesives. But skipping the initiation phase will not significantly change the overall fatigue results if the life is dominated by the crack propagation phase. It is important to be aware that adhesive nature can have an important effect on this behavior and joints bonded with more ductile adhesives will often experience a longer crack propagation life compared to the initiation stage.

1.4 High Cycle Fatigue (HCF) Versus Low Cycle Fatigue (LCF)

The fatigue response of adhesive joints can be divided into high cycle fatigue and low cycle fatigue regimes. Some authors believe that for joints where the maximum fatigue load applied is greater than or equal to the yield strength of the adhesive, the joint will have a low fatigue life cycle. Consequently, for a high cycle fatigue regime, the load level must be set below the yield point of the tested material. Another criterion for separating LCF and HCF is the total number of fatigue lives. Typically, a total fatigue life of less than a thousand cycles is considered to be LCF. Based on this definition for LCF, the adhesive undergoes cyclic plastic deformation. However, it should be noted that this definition is primarily valid for load control conditions.

The crack initiation and propagation life discussed in the previous section are in some way related to the LCF and HCF concepts. For the LCF regime, life is mainly governed by crack propagation and for the HCF condition the initiation phase may be dominant. By reducing the applied load, the joint will reach a strength condition where no failure is observed in very high cycle fatigue, which is generally defined to consist of at least 1 million cycles. Figure 1.1 shows three different regions (LCF, HCF, and the endurance conditions). Endurance conditions are shown by stress levels at which the fatigue life of the joints can be in practice considered as infinite.

Fig. 1.1 LCF, HCF, and endurance limit regions

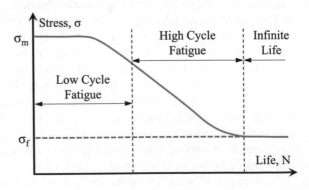

1.5 Fatigue Characterization

Adhesives can be experimentally characterized in terms of fatigue, following two different test procedures. The first approach characterizes adhesives in terms of total fatigue life (sometimes called fatigue life initiation). The second procedure deals with the analysis of fatigue crack growth of adhesives. Different joint geometries have been used by researchers to analyze the total fatigue life of adhesives. Single lap joints, scarf joints, double lap joints, Arcan joints, and T-joints are the most common joint configurations used for the fatigue analysis of adhesives. Using Arcan specimens (see Fig. 1.2), different mode mixities from pure mode I to pure mode II can be created simply by rotating the Arcan device. Different levels of the load must be applied to the joints to construct the stress-life (S–N) curves as a function of the mode ratio.

By considering an appropriate equivalent stress parameter, the stress life curves obtained for different loading conditions can be collected into a single master S–N curve, which defines the total fatigue response of the tested adhesive. However, this approach is primarily used for joints where fatigue life is dominated by the onset of damage. In contrast to the S–N characterization, fatigue crack growth analysis of adhesives is a more complex and challenging activity. Fatigue crack growth data

Fig. 1.2 An Arcan assembly

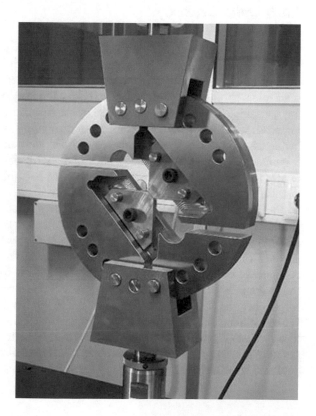

of adhesives is often required for more advanced designs, those where the bonded joints are used in critical load-bearing components. Standardized fracture specimens such as double cantilever beam (DCB) and end notched flexure (ENF) are often considered, respectively, in conditions of mode I and mode II to characterize the adhesives in terms of fatigue crack propagation. From these tests, not only the crack growth rate but also the fatigue limit parameters can be obtained. One of the most challenging parts of these tests is the measurement of the crack length as a function of load cycles. Different techniques are available for this purpose, such as manual measurement of the crack or the use of digital image correlation (DIC) techniques. However, one of the most practical and effective methods is the compliance-based approach, where the load line displacement measured by the testing machine during the fatigue test is employed to estimate the crack size. One of the most well-known compliance methods is the compliance-based beam method (CBBM), which has been effectively used by several authors [9–11] to measure the crack size during fatigue testing of DCB and ENF samples.

1.6   Factors Affecting the Fatigue of Adhesives

1.6.1   Joint Geometry

Similarly, to what occurs for static strength, the fatigue response of joints is mainly a function of joint geometry. Through a literature review, it has been found that different types of joints have been considered for fatigue life assessment of adhesives. Various scarf joints geometries were fatigue loaded by Jen [12] to determine the influence of the scarf angle on the fatigue life of the joints, discovering that the total fatigue life increases with the angle of the scarf. In another study, two different types of joints, including double lap joints and stepped joints, were analyzed in terms of fatigue response [13]. The fatigue performance of butterfly joints was analyzed by Altan et al. [14], concluding that, when compared to butt joints, the butterfly geometry leads to longer life cycles. The effect of the overlap area on fatigue life was analyzed by da Costa Mattos et al. [15] by fatigue loading of adhesively bonded composite single lap joints. Although larger overlaps are generally thought to lead to better static strength and thus increase fatigue performance, the results reported by the authors show that for specific loads and material conditions, fatigue life may decrease with an increase of the overlap length which may be due to increased peel stresses at the overlap ends. Adhesive bonded sheet molding composites were analyzed by Mazumdar and Mallick [16]. In contrast to Jen and Ko [17], Mazumdar and Mallick [16] concluded that overlap does not significantly change the life of single lap joints. Their results showed that it is the adhesive thickness that plays an important role in the fatigue response of the tested samples. The effect of adhesive thickness was also analyzed by Azari et al. [18, 19].

1.6.2   Effect of Material Property

Generally speaking, the higher the brittleness of the adhesive the shorter the fatigue crack propagation life is. In fact, for the more brittle materials the total fatigue life is mainly governed by the fatigue damage initiation. On the other hand, joints with ductile adhesives may experience a longer fatigue crack propagation life. Nowadays, a new generation of adhesives is available in the market that is able to combine high strength, large ductility and plastic deformation before failure, leading to excellent fatigue performance.

1.6.3   Effect of Loading Conditions

Several studies have been conducted on the effect of loading conditions on the fatigue performance of bonded structures. The R-ratio (the ratio of minimum load or stress to maximum load or stress), loading frequency, average stress, maximum load, mode mixity, and stress amplitude are the main loading factors that influence the fatigue behavior of adhesive materials. Figure 1.3 shows a typical constant amplitude sinusoidal fatigue load spectrum. The relationship between the different loading parameters is also shown in Eqs. 1.1–1.4.

$$\Delta\sigma = \sigma_{max} - \sigma_{min} \tag{1.1}$$

$$\sigma_a = \frac{\sigma_{max} - \sigma_{min}}{2} \tag{1.2}$$

$$\sigma_m = \frac{\sigma_{max} + \sigma_{min}}{2} \tag{1.3}$$

$$R = \frac{\sigma_{min}}{\sigma_{max}} \tag{1.4}$$

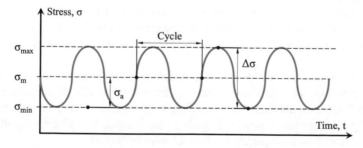

Fig. 1.3   A typical sinusoidal fatigue loading

Fig. 1.4 Schematic of a variable amplitude fatigue loading

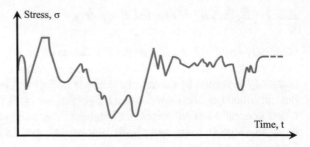

Although in most studies joints are subjected to a constant amplitude loading condition, in real applications, the amplitude of the service load is in fact not constant and does change with time. This type of fatigue loading is called variable amplitude loading condition (see Fig. 1.4).

Variable amplitude loading can either improve the fatigue life of joints by retarding crack growth or reduce fatigue life by accelerating crack growth. Both phenomena are caused by overloads applied under varying amplitude loading conditions. Compressive residual stresses in the vicinity of the crack tip, crack tip closing effects and crack tip blunting are the main mechanisms thought to retard fatigue crack growth in joints subjected to variable amplitude loading conditions [20].

1.6.4 Environmental Effects

A combination of aggressive environments and fatigue loading can significantly reduce the service life of bonded joints. Plasticization is the main phenomenon that occurs when a joint is subjected to conditions of humidity or high temperature that can reduce the rigidity and strength of adhesives. However, plasticization can also lead to a greater plastic deformation of adhesives. Since higher ductility will lower the stress concentration factors, plasticization can sometimes reduce the importance of early life fatigue damage in the overall fatigue life of joints.

Another damage mechanism associated to environmental effects are the residual stresses induced by ageing and by high temperatures. High temperatures and the swelling caused by water absorption can introduce significant residual stresses within the bondline which are often damaging and can lead to the premature fatigue failure of bonded joints.

Water absorption not only degrades the adhesive but it can also reduce the mechanical properties of the adhesive-adherend interface, especially at the edges of the bondline. Interfacial ageing at these locations leads to adhesive failure in these areas, increasing increase the stresses acting along the overlap by reducing the effective bonded area which consequently leads to a lower fatigue life.

High temperature working conditions, especially when the service temperature is close to the glass transition temperature of the adhesive, can also change the behaviour of joints subjected to cyclic loads. Figure 1.5a shows the effect of temperature and humidity on the S–N behaviour of adhesives. The effects of humidity and temperature on the crack growth (da/dN) is also shown in Fig. 1.5b. Where a is the crack length, N is the cycle number and G is the strain energy release rate calculated for each loading cycle.

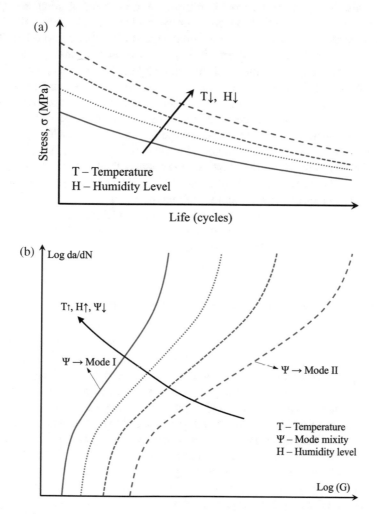

Fig. 1.5 a Effects of temperature and humidity on the S–N curves and **b** effects of mode mixity, temperature and humidity on the Paris law curve

1.7 Fatigue Threshold Energy

The value of the fatigue threshold is required for some fatigue life analysis approaches, since threshold values determine the load level at which fatigue life is sufficient for a reliable service. Based on the proposed definitions, below the fatigue limit, there is no crack growth or the crack growth rate is small enough for attaining a fatigue life of 1 million cycles or more. However, measuring the fatigue threshold energy of adhesives is a great challenge since, similarly to other fatigue properties of adhesives, this property is a function of joint geometry and loading mode. The mode mixity, geometry of the joint, and the test approach can all influence the fatigue threshold obtained from the tests. Two different methods can be used to obtain this fatigue parameter, including the load control approach and the displacement control method. For both conditions, the test can be conducted on a specimen with an initial pre-crack or it can be performed on a joint without any initial crack. Some authors have already analyzed the values of fatigue thresholds obtained by different techniques [21, 22]. Based on the results of Johnson and Mall [23], no significant difference was found between the fatigue threshold energy obtained from pre-cracked samples and from samples without initial cracks.

Fatigue testing can be carried out with load control and displacement control tests. However, there are some important differences between these two testing strategies, discussed in detail in the following section.

1.8 Force Versus Displacement Control Tests

As mentioned above, two different fatigue test methodologies can be adopted to characterize the fatigue behavior of adhesives. The first approach is load control, where the maximum and minimum fatigue loads applied are precisely defined. In this type of test, the crack growth rate increases with loading cycles. Since the applied load is constant, the increase in crack size will increase the strain energy level, which eventually leads to an increase in the fatigue crack growth rate. The strain energy release rate will finally reach a critical value where the crack grows unsteadily leading to the joint failure. On the other hand, in displacement control tests, the initial crack growth is faster, but it reduces cycle by cycle. Once displacement is controlled, the load (and crack growth rate) must be constantly reduced to a maintain constant displacement. By reducing the applied load one can control the level of strain energy release rate introduced. Eventually, the energy level will be reduced to a value below the threshold energy where no fatigue crack grows. Thus, using the displacement control technique, fatigue fracture of joints can be better controlled. Interestingly, some authors believe that both load control and displacement control give similar results [24, 25], but displacement control can also give the possibility to test different conditions using a single sample.

1.9   Life Prediction Approaches

Having the adhesive characterized in terms of fatigue behavior is not sufficient for evaluating the life of real bonded structures, since this data must be fed to suitable fatigue life prediction models, supported by finite element modelling and based on fatigue damage onset or crack growth concepts. In terms of fatigue initiation life, S–N based techniques are the most popular methods used for life assessment of joints subjected to cyclic loading. However, techniques based on fracture mechanics concepts are also employed to investigate fatigue crack growth in adhesive materials.

1.9.1   Total Fatigue Life (S–N) Approaches

When the total fatigue life is mainly governed by the fatigue damage initiation phase, the initial life is generally considered to be the total fatigue life of the component. This assumption will result in a more conservative life, leading to a safer joint design. The main tool needed to estimate service life with S–N techniques is a master S–N curve where an equivalent stress parameter is plotted against the life of the joint.

A combination of this master curve with the finite element method can be employed to predict the fatigue life of bonded structures, with the equivalent parameter being usually based on the stress components. However, some authors have developed strain-based models or a combination of strain- and stress-based methods [26–29]. FEM can also be based on the classical linear elastic assumption or can be further developed to include the plasticity of the materials tested. However, it should be noted that the total life approach is best suited for high cycle fatigue regimes, where the stress level is well below the yield limit of the adhesive. Figure 1.6 schematically shows the fatigue life estimation procedure for bonded joints using the S–N approach. More details can be found in [30].

1.9.2   Fatigue Crack Growth Method

While in the early stages of joint design an S–N approach can be useful but for advanced joint designs, especially for lightweight structures, precise determination of the crack growth is still of paramount importance. Taking crack propagation life into account in joint design can lead to a safer and longer service life and lower structural weight. The analysis of fatigue crack growth is based mainly on the concepts of linear elastic fracture mechanics. In this approach, the crack growth rate is measured as a function of an equivalent strain energy release rate (G) or stress intensity factor (K). Several equivalent parameters were defined by the authors. In [31] the authors reviewed and compared the most important energy-based models. The equivalent strain energy release rate must take into account the effects of different loading

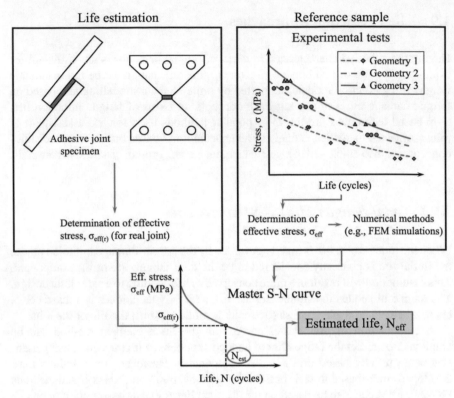

Fig. 1.6 S–N procedure used for fatigue life estimation of adhesive joints

conditions and the crack growth rate, as a function of the equivalent G(K), is plotted in a log–log graph known as the Paris law curve. The Paris law curve thus consists of the three regions shown in Fig. 1.7. Equation 1.5 also shows a typical Paris law relationship based on the rate of energy release (G).

$$\frac{\mathrm{d}a}{\mathrm{d}N} = f(G) \tag{1.5}$$

where the rate of crack growth is given by $\mathrm{d}a/\mathrm{d}N$. Different relations have been defined for $f(G)$ for example $f(G) = C(G)^m$ where C and m are fitting parameters. Based on the crack growth rate approach, the initial crack size (a_0) and the critical length of the crack (a_c) should be previously known, which allows to reorder Eq. 1.5 and estimating life using the following relation.

$$N = \int_{a_0}^{a_c} \frac{\mathrm{d}a}{f(G)} \tag{1.6}$$

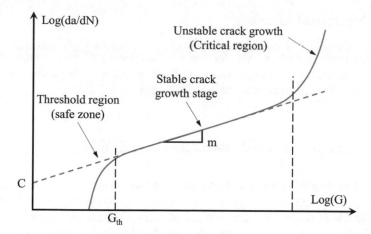

Fig. 1.7 A typical Paris law curve

where a_0 is the initial crack size and ac the critical crack length.

Ignoring the damage initiation phase may cause an underestimation of fatigue life might be leading to increasing the costs and weight of the structure. Accordingly, the crack growth rate approach is mostly suitable for joints in which a significant part of the total fatigue life is spent in the crack propagation stage.

1.9.3 Damage Models

Due to the limitations of the S–N and FCG approaches, new methods have been recently developed, based on damage models, which can simulate the damage initiation and propagation phases of the fatigue life of bonded joints. These methods simulate damage within the adhesive layer in an incremental process. Different fatigue damage models have been proposed by the authors including continuum damage mechanics and models based on fracture mechanics concepts [32–36]. Based on the damage analysis models, a damage parameter called D is defined, governing the variation in the properties of the material. Applying the damage value in each load cycle, the adhesive properties are updated (degraded) in each step of the numerical analysis, reducing joint strength until ultimate failure. More information on numerical methods is provided in the following section.

1.10   Numerical Methods

Many different approaches exist for the implementation of the damage analysis models in numerical analysis programs. In this section, we will briefly discuss the most commonly used numerical analysis techniques.

1.10.1   Extended Finite Element Method, XFEM

XFEM is based on the concept of the partition unit where the sum of shape functions equals one. Using XFEM it is possible to define discontinuities within the model, solving problems where the damage initiation point is not known a priori. Crucially, using XFEM, a model will not need to be remodeled as the crack grows through the material. Using the XFEM approach, the path where the crack grows does not need to be known or defined first. Recently, XFEM has been used for fatigue analysis of specimens [37–40]. A combination of XFEM and the cohesive zone (CZ) approach is also advanced in some studies [41, 42]. However, more research is still needed to assess its performance and capabilities under cyclic loading conditions. Furthermore, there are still some challenges associated to implementing XFEM in commercial finite element packages and XFEM for 3D conditions is only available for solid elements and for planar models when 2D conditions are considered.

1.10.2   Interfacial Thick Level Set Modeling

Another method which can be used to simulate damage within the bondline is the adoption of the interfacial thick level set, an approach initially introduced by Moës [43]. Based on this approach, a specific length called damage zone is defined and the energy release rate is calculated along this zone by an integration technique. This approach uses the same base concept as XFEM and CZM methods (discussed in the next subsections) but differs from these in the fact that damage is not a function of displacement but it is instead a function of a level set field defined at the interface or at the bondline. Latifi et al. [44] used the interfacial thick level set modelling for analysis of fatigue initiation and propagation in composites. In this approach, the shape of the profile of the damage as well as the damage length (that is, the same as the length of the fracture process zone FPZ) should be known a priori. A comparison between the thick level set approach and CZM was made by Gomez et al. [45]. The authors report that the thick level set can simulate the damage as precisely as the CZM method.

1.10.3 CZM

Fatigue life analysis, using the concepts of linear elastic fracture mechanics (LEFM), has some important limitations such as the need for the presence of an initial flaw within the bonding line, ignoring the fatigue initiation life, and not being able to consider the effect of the non-linear behavior of the materials. To overcome these issues, a robust criterion based on the constitutive modeling of the process zone (cohesive zone) was proposed. CZM has been widely used to analyze the strength of bonded joints. In CZM, the traction on cohesive elements is defined as a function of separation and a damage parameter. The damage parameter is defined to measure the degradation of the properties of the cohesive element as a function of the separation level. The CZM concept was proposed by Dugdale in 1960 and Barenblatt in 1962 [46]. This technique was later combined with FEM by Hillerborg et al. (1976) [47]. Today, CZM is a well-known technique, adopted by researchers and industry alike to simulate the mechanical response of bonded joints subjected to various loading conditions, including cyclic loading.

The main aim of the current book is to explain in detail how CZM is used for fatigue life assessment of adhesives. Accordingly, after a brief review of the fatigue in adhesive joints and introducing different analysis approaches in the current chapter, Chap. 2 will talk about the CZM in greater detail, discussing different cohesive laws and their definition. In Chap. 2, the constitutive laws of the cohesive elements for different mode mixities are also presented. In Chap. 3, the CZM-based approaches used for fatigue analysis of adhesive joints are discussed. Different degradation methods applied to the cohesive properties proposed by the authors and different strategies used in the fatigue simulation of bonded joints using CZM are explained. In Chap. 4 the steps and approaches in numerical simulation of fatigue are explained. In this chapter, it is discussed how CZM fatigue models can be implemented in commercial FE software. Finally, Chap. 5 summarizes the book contents and presents the conclusions.

References

1. Dennis, W.H. 2017. *Metallurgy*, 1863–1963. Routledge.
2. Jones, R., et al. 2020. Requirements and variability affecting the durability of bonded joints. *Materials* 13 (6): 1468.
3. Raizenne, D. 2002. Case history: CF-116 upper wing skin fatigue enhancement boron doubler. In *Advances in the Bonded Composite Repair of Metallic Aircraft Structure*, 937–957. Elsevier.
4. Bidaud, P. 2014. *Analysis of the Cyclic Behavior of an Adhesive in an Assembly for Offshore Windmills Applications*. Université de Bretagne occidentale-Brest.
5. Imanaka, M., K. Haraga, and T. Nishikawa. 1993. Fatigue strength of adhesive/rivet combined joints. *Adhesion'93* 187–192.
6. Harris, J., and P. Fay. 1992. Fatigue life evaluation of structural adhesives for automative applications. *International Journal of Adhesion and Adhesives* 12 (1): 9–18.

7. Sousa, F.C., et al. 2020. The influence of mode mixity and adhesive system on the fatigue life of adhesive joints. *Fatigue & Fracture of Engineering Materials & Structures* 43 (10): 2337–2348.

8. Lefebvre, D., D. Dillard, and J. Dillard. 1999. A stress singularity approach for the prediction of fatigue crack initiation in adhesive bonds. Part 2: Experimental. *The Journal of Adhesion* 70 (1–2): 139–154.

9. Rocha, A., et al. 2020. Numerical analysis of mixed-mode fatigue crack growth of adhesive joints using CZM. *Theoretical and Applied Fracture Mechanics* 102493.

10. Rocha, A.V., et al. 2020. Fatigue crack growth analysis of different adhesive systems: Effects of mode mixity and load level. *Fatigue & Fracture of Engineering Materials & Structures* 43 (2): 330–341.

11. Monteiro, J., et al. 2020. Influence of mode mixity and loading conditions on the fatigue crack growth behaviour of an epoxy adhesive. *Fatigue & Fracture of Engineering Materials & Structures* 43 (2): 308–316.

12. Jen, Y.-M. 2012. Fatigue life evaluation of adhesively bonded scarf joints. *International Journal of Fatigue* 36 (1): 30–39.

13. Zhang, Y., A.P. Vassilopoulos, and T. Keller. 2008. Stiffness degradation and fatigue life prediction of adhesively-bonded joints for fiber-reinforced polymer composites. *International Journal of Fatigue* 30 (10–11): 1813–1820.

14. Altan, G., M Topçu, and H Çallıoğlu. 2010. The effects of the butterfly joints on failure loads and fatigue performance of composite structures. *Science and Engineering of Composite Materials* 17 (3): 199–212.

15. da Costa Mattos, H., A. Monteiro, and R. Palazzetti. 2012. Failure analysis of adhesively bonded joints in composite materials. *Materials & Design* 33: 242–247.

16. Mazumdar, S., and P. Mallick. 1998. Static and fatigue behavior of adhesive joints in SMC-SMC composites. *Polymer Composites* 19 (2): 139–146.

17. Jen, Y.-M., and C.-W. Ko. 2010. Evaluation of fatigue life of adhesively bonded aluminum single-lap joints using interfacial parameters. *International Journal of Fatigue* 32 (2): 330–340.

18. Azari, S., M. Papini, and J. Spelt. 2011. Effect of adhesive thickness on fatigue and fracture of toughened epoxy joints–Part I: Experiments. *Engineering Fracture Mechanics* 78 (1): 153–162.

19. Azari, S., M. Papini, and J. Spelt. 2011. Effect of adhesive thickness on fatigue and fracture of toughened epoxy joints–Part II: Analysis and finite element modeling. *Engineering Fracture Mechanics* 78 (1): 138–152.

20. da Silva, L.F.M., A. Öchsner, and R.D. Adams. 2018. *Handbook of Adhesion Technology*, vol. 1. Springer.

21. Ashcroft, I., and S. Shaw. 2002. Mode I fracture of epoxy bonded composite joints 2. Fatigue loading. *International Journal of Adhesion and Adhesives* 22 (2): 151–167.

22. Jethwa, J., and A. Kinloch. 1997. The fatigue and durability behaviour of automotive adhesives. Part I: fracture mechanics tests. *The Journal of Adhesion* 61 (1–4): 71–95.

23. Johnson, W., and S. Mall. 1985. A fracture mechanics approach for designing adhesively bonded joints. In *Delamination and Debonding of Materials*. ASTM International.

24. Mall, S., G. Ramamurthy, and M. Rezaizdeh. 1987. Stress ratio effect on cyclic debonding in adhesively bonded composite joints. *Composite Structures* 8 (1): 31–45.

25. Azari, S., et al. 2010. Fatigue threshold behavior of adhesive joints. *International Journal of Adhesion and Adhesives* 30 (3): 145–159.

26. Harris, J., and R. Adams. 1984. Strength prediction of bonded single lap joints by non-linear finite element methods. *International Journal of Adhesion and Adhesives* 4 (2): 65–78.

27. Bigwood, D., and A. Crocombe. 1990. Non-linear adhesive bonded joint design analyses. *International Journal of Adhesion and Adhesives* 10 (1): 31–41.

28. Goglio, L., M. Rossetto, and E. Dragoni. 2008. Design of adhesive joints based on peak elastic stresses. *International Journal of Adhesion and Adhesives* 28 (8): 427–435.

29. da Silva, L.F., et al. 2009. Analytical models of adhesively bonded joints—Part II: Comparative study. *International Journal of Adhesion and Adhesives* 29 (3): 331–341.

30. Castro Sousa, F., et al. 2020. Fatigue life estimation of adhesive joints at different mode mixities. *The Journal of Adhesion*, 1–23.
31. Rocha, A., et al. 2019. Paris law relations for an epoxy-based adhesive. *Proceedings of the Institution of Mechanical Engineers, Part L: Journal of Materials: Design and Applications* 1464420719886469.
32. Wahab, M.A., et al. 2001. Prediction of fatigue thresholds in adhesively bonded joints using damage mechanics and fracture mechanics. *Journal of Adhesion Science and Technology* 15 (7): 763–781.
33. Imanaka, M., et al. 2003. Fatigue damage evaluation of adhesively bonded butt joints with a rubber-modified epoxy adhesive. *Journal of Adhesion Science and Technology* 17 (7): 981–994.
34. Hilmy, I., et al. 2007. Effect of triaxiality on damage parameters in adhesive. In *Key Engineering Materials*. Trans Tech Publications Ltd.
35. Sheppard, A., D. Kelly, and L. Tong. 1998. A damage zone model for the failure analysis of adhesively bonded joints. *International Journal of Adhesion and Adhesives* 18 (6): 385–400.
36. Edlund, U., and A. Klarbring. 1993. A coupled elastic-plastic damage model for rubber-modified epoxy adhesives. *International Journal of Solids and Structures* 30 (19): 2693–2708.
37. Singh, I., et al. 2012. The numerical simulation of fatigue crack growth using extended finite element method. *International Journal of Fatigue* 36 (1): 109–119.
38. Pathak, H., A. Singh, and I.V. Singh. 2013. Fatigue crack growth simulations of 3-D problems using XFEM. *International Journal of Mechanical Sciences* 76: 112–131.
39. Sabsabi, M., E. Giner, and F. Fuenmayor. 2011. Experimental fatigue testing of a fretting complete contact and numerical life correlation using X-FEM. *International Journal of Fatigue* 33 (6): 811–822.
40. Xu, Y., and H. Yuan. 2009. Computational analysis of mixed-mode fatigue crack growth in quasi-brittle materials using extended finite element methods. *Engineering Fracture Mechanics* 76 (2): 165–181.
41. Wells, G.N., and L. Sluys. 2001. A new method for modelling cohesive cracks using finite elements. *International Journal for Numerical Methods in Engineering* 50 (12): 2667–2682.
42. Mergheim, J., E. Kuhl, and P. Steinmann. 2005. A finite element method for the computational modelling of cohesive cracks. *International Journal for Numerical Methods in Engineering* 63 (2): 276–289.
43. Moës, N., et al. 2011. A level set based model for damage growth: The thick level set approach. *International Journal for Numerical Methods in Engineering* 86 (3): 358–380.
44. Latifi, M., F. van der Meer, and L. Sluys. 2017. An interface thick level set model for simulating delamination in composites. *International Journal for Numerical Methods in Engineering* 111 (4): 303–324.
45. Gómez, A.P., N. Moës, and C. Stolz. 2015. Comparison between thick level set (TLS) and cohesive zone models. *Advanced Modeling and Simulation in Engineering Sciences* 2 (1): 1–22.
46. De Moura, M., and J. Gonçalves. 2014. Cohesive zone model for high-cycle fatigue of adhesively bonded joints under mode I loading. *International Journal of Solids and Structures* 51 (5): 1123–1131.
47. Hillerborg, A., M. Modéer, and P.-E. Petersson. 1976. Analysis of crack formation and crack growth in concrete by means of fracture mechanics and finite elements. *Cement and Concrete Research* 6 (6): 773–781.

Chapter 2
Cohesive Zone Modelling-CZM

Abstract Experimental fatigue tests are very expensive and time-consuming procedures. Accordingly, numerical techniques are often employed to better understand
the fatigue behavior of bonded joints and to predict their fatigue life. Several numerical techniques based on continuum mechanics, fracture mechanics and damage
mechanics concepts have been proposed to estimate the fatigue response of bonded
structures. Methods based on the continuum mechanics mainly focus on the initiation fatigue life while approaches based on the fracture mechanics concepts consider
fatigue crack growth. But damage mechanics concepts can simulate both the fatigue
life initiation and the fatigue crack propagation. Accordingly, damage mechanics
approaches have recently started to attract more attention. One of the most popular
of these approaches is known as cohesive zone modelling (CZM). Various types
of CZMs have been proposed so far but considering the variety of the developed
CZM shapes, a question that arises here is which CZM shape is more suitable for
the joint analysis, a question which is answered in this chapter. Furthermore, the
properties of the most common traction separation law (TSL) mentioned above are
also reviewed. However, the next question is how different parameters in the considered TSL can be identified. Different techniques to shape a TSL are reviewed in this
chapter including the classical methods, inverse technique, and direct approach. The
last of these approaches is the most precise one since for the first two approaches
the shape of the CZM should be selected beforehand. In this chapter, the constitutive
laws of different CZMs are also presented and discussed. Using the constitutive laws,
the cohesive models can be implemented in FE programs.

2.1 History and Concepts

As mentioned in Chap. 1, the CZM technique is widely used for damage analysis of
bonded joints, driven by its important advantages, such as less sensitivity to the mesh
configuration, the lack of an initial defect, the ability to simulate both the initiation
and propagation fatigue life and flexibility to take into account the effects of different
loading and environmental conditions. The CZM concepts were first introduced by

Dugdale [1] and Barenblatt [2]. Dugdale considered the plastic deformation of the materials ahead of the crack tip. However, his model was used for materials with small-scale flow behavior [3] and with a perfect linear elastic plastic response. Using Dugdale's approach, the singularity problems at the tip of the cracks were solved. Figure 2.1a schematically shows the Dugdale model.

Barenblatt extended the work of Dugdale by replacing the perfect plastic behaviour with a stress function where the stress is a function of the displacement of the crack tip (Fig. 2.1b). Similar to the Dugdale model, the Barenblatt approach avoids the singularity of the stress at crack tip. Using the Barenblat method, the response of the specimen to the applied load is a function of the stress function defined for the process zone ahead of the crack tip. Since then, the use of CZM has constantly grown, finding place in both the industrial and research sectors. To define the behavior of adhesives using CZM, different stress functions for the process zone can be defined. One of the first and simplest ones was introduced by Hiller-borg et al. [4], which proposed a linear relationship between the traction (stress)

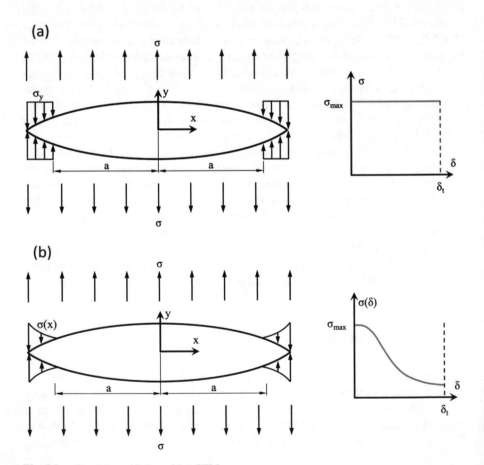

Fig. 2.1 a Dugdale and **b** Barenblatt CZM

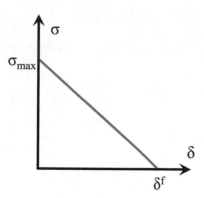

Fig. 2.2 Hillerborg CZM model

and separation (displacement) of the material in the process zone. Figure 2.2 shows Hillerborg et al. model. Due to its simplicity, this stress function is widely used, since its implementation in finite element methods is quite easy.

At this stage, it is important to note that CZM does not directly show the physical behavior of materials. However, using CZM it is possible to take into account the stresses applied to the material points in front of the crack tip. In CZM, the traction separation law (TSL) relates the material separation to the applied stress. The area under the TSL is called fracture energy, which is the energy required to cause a complete separation of material points. Each material has its own ideal CZM shape, which best represents its specificities. There are three different approaches to define the TSL for a material, discussed in detail later in Sect. 2.3. There are certain parameters in each TSL that define the shape of the CZM. One is the fracture energy, the other is the separation at which the damage initiates, and the third important parameter is the maximum separation at which a complete failure occurs. These points are discussed later in this chapter in Sect. 2.2. Based on the type of the material and loading conditions, different TSLs have been defined by the authors and the next section discusses the most common TSL shapes that have been introduced.

2.2 Traction Separation Laws

Generally, a CZM traction separation law consists of an initial increase in traction until a maximum point is attained, followed by a damage evolution section where the traction decreases and damage increases with added displacement.

Figure 2.3 shows a typical bilinear TSL. According to the CZM shape shown in Fig. 2.3, when the separation (displacement) reaches a specific value (called damage initiation point) the damage initiates. Before this point the material will not experience any damage, which means that by unloading the specimens the cohesive properties of the material will return to their initial conditions. However, any further increase in displacement above the initiation point will cause and increase the damage. By increasing the displacement, the maximum traction load that the material can sustain

Fig. 2.3 A typical bilinear
TSL

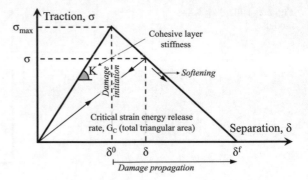

decreases. A maximum separation is defined in TSLs at which a complete separation takes place. At this point the damage value reaches 1 which means that at the considered material point a discontinuity occurs and the material will not sustain any further load at that point.

Although following similar concepts, distinct forms of the TSLs have been proposed to fit better with the behaviour of materials. The extensive studies on the shape of the CZM and the various TSLs developed by authors are mainly due to the difference in the mechanical properties of the materials. The most common shapes of the TSL are triangular (bilinear), trapezoidal, exponential, polynomial, trilinear, and linear parabolic, introduced by diverse researchers [5–9]. These are shown in Fig. 2.4. Although some authors believe that the shape of the CZM does not play a key role on the response of the materials [8], others have carried out extensive studies that support the idea that the shape of the TSL is indeed an important factor in a precise damage simulation [5, 10–13]. In the next section, the most widely used forms of the TSL are discussed and the corresponding constitutive laws are presented.

2.2.1 Bilinear TSL

One of the simplest TSLs is the single line (is usually classified as bilinear) CZM form shown in Fig. 2.2. Hillerboorg [4] first developed this model for use with brittle materials. The high initial stiffness in this approach introduces difficulties in numerical implementations. However, for less brittle materials, the bilinear shape was introduced with success. As pointed out in the previous section, in Bilinear TSLs the cohesive response of the adhesive in both the initial part and for the damage evolution section is considered as linear. In this initial part the slope of the line represents the stiffness of the cohesive elements which is a penalty parameter for interfacial cohesive elements. A wide range of values can be considered as the penalty stiffness. However, as it will be discussed later in Chap. 3, for finite thicknesses of cohesive layers the initial stiffness can be considered as the Young's modulus of the material, divided by the thickness of the cohesive elements. In Bilinear TSL, by reaching the maximum

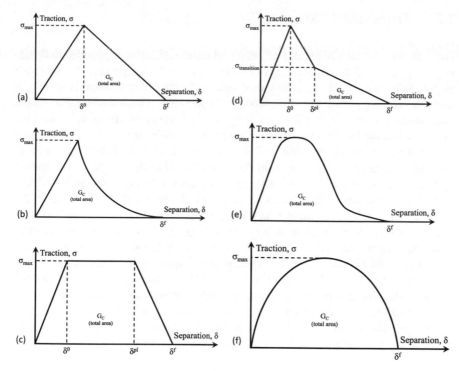

Fig. 2.4 Some of the most common CZM shapes **a** bilinear, **b** linear exponential, **c** trapezoidal, **d** trilinear, **e** polynomial, **f** parabolic

traction value, the damage initiates and any further increase in displacement will increase the damage level until its value reaches 1, where a complete failure occurs. The maximum traction is thus a key parameter in all the TSLs, as this is also the case for the bilinear shape. By knowing the initial stiffness, the maximum traction, and the fracture energy, the bilinear TSL can be precisely defined. Experimental/numerical techniques that are employed to obtain these parameters are discussed in detail in Sect. 2.3.

In the damage evolution section, the stiffness of the material is degraded. It is evident that by keeping all parameters constant, any reduction in initial stiffness will increase the displacement at which damage starts resulting in a small part of the damage evolution.

Due to its simplicity, bilinear TSL is almost available in the most common FE software, e.g. Abaqus. A bilinear TSL is easily formulated and can provide accurate results for brittle adhesives where no significant plasticity is observed after material flow. We call the attention of the author to the fact that some authors believe that the bilinear model can be considered a special case of the trapezoidal TSL, discussed in the next section.

2.2.2   Trapezoidal TSL

For more ductile materials, especially those where the adhesive experiences a significant FPZ ahead of the crack tip, the bilinear shape might not properly simulate the response of the joint. In this case a modification to the bilinear shape is needed to take the adhesive plastic flow into account. To this end, a trapezoidal CZM shape, schematically shown in Fig. 2.4, was proposed by Tvergaard and Hutchinson [14] that considers the behaviour of adhesives with larger FPZ size.

In contrast to the simple bilinear TSL, the trapezoidal TSL shape requires more material parameters to be defined in order to determine the length of the central plateau of the trapezoidal model. Regardless of the values set to shape the middle part of the trapezoidal CZM, the area under the TSL should always be equal to the fracture energy of the considered adhesive, a parameter that is usually experimentally obtained. Except for the fracture energy, the point at which the damage initiates must be defined in a trapezoidal TSL. As shown in Fig. 2.4 in the second zone the traction is kept constant by increasing the displacement. However, it should be noted that although no degradation in the traction level is defined, damage in the material does increase with this displacement. Tvergaard and Hutchinson [14] showed that this newly introduced parameter has a relatively reduced influence on the results, but generally speaking, the sensitivity of the results to this parameter is known to depend on the loading conditions and the type of the tested material.

2.2.3   Linear-Exponential TSL

The exponential TSL was introduced [15–17] and soon modified by several authors to consider the behaviour of the different material behaviors. This CZM law was mainly developed for shear loading conditions. One of the variant forms of the exponential TSL is the linear-exponential form where the initial part of the TSL is considered as linear up to the damage initiation point. In the exponential TSL the damage evolution part is exponentially related to the separation. Compared to the bilinear shape, a higher drop in traction is observed right after the damage initiation, changing the response of the cohesive elements when they reach the maximum traction. In some studies, it is shown that the exponential TSL is quite an appropriate model for materials with significant plastic behaviour before failure.

2.2.4   Multi-linear Softening TSL

Although this is not a commonly used shape, some authors have employed multi-linear softening TSL forms to model adhesives. In multi linear softening CZM, the initial part is the same as the bilinear shape where the traction linearly increases

Fig. 2.5 Bilinear damage evolution

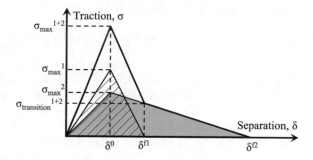

until the maximum traction. However, instead of using a single line to define the damage evolution part, several lines with different slopes and lengths are used to form the softening part of the TSL. Using multiple lines to define the damage evolution behaviour increases model complexity and cause difficulties in implementation in FE programs. However, in some works it has been shown that the additional effort needed by this approach can lead to more precise results for some material and geometry conditions. Trapezoidal TSL discussed above is a specific case of the multi-linear TSLs.

A trilinear CZM (two linear softening behaviour) was proposed by Heidari-Rarani and Ghasemi [11] as schematically shown in Fig. 2.5. They found a better agreement with the experimental data by using the bilinear softening TSL compared to the single line softening CZM. The proposed CZM introduced by Heidari-Rarani and Ghasemi [11] was constructed by superimposing two bilinear TSLs.

The main goal of using multi-linear softening behaviour is to take into account different failure mechanisms that occurs in the material, especially in composites. The authors [11] also state that the implementation of their proposed trilinear model in FE programs is much simpler.

2.2.5 Customized TSL

In addition to the well-defined material models discussed above, some researchers believe that the TSL must be obtained directly from specific experimental tests. Consequently, the way these TSLs are obtained is called "direct method". The shape of custom TSLs is generally of polynomial nature. However, no specific relationship can be typically found for the initial damage free section nor for the softening part. Custom or direct TSLs generally provide more accurate results compared to other common forms. However, obtaining the TSL form and implementing it in commercial FE programs is an intricate process, which requires highly specific knowledge. The direct approach to measuring TSL parameters is discussed in Sect. 2.3.

2.2.6   Which TSL Should Be Used?

As discussed above, several different shapes have been proposed to account for
the variety of the behaviour of different adhesives. Some studies have numerically
analyzed the performance of different TSLs and compared it with experimental
results. Most of these works focus on the bilinear, trapezoidal, and exponential
TSLs. The bilinear TSL was compared with the exponential model proposed by
Chandra et al. [8]. They applied these models to the interface of titanium-matrix
composites, reinforced with SCS (silicon carbide) fibers. They found that, for the
materials tested, the bilinear CZM can accurately reproduce the load displacement
response, whereas the exponential TSL cannot do so. Alfonso [5] analyzed different
TSL shapes including bilinear, exponential, linear-parabolic, and trapezoidal. The
authors applied this model to pure-mode loading conditions and found that the ratio
between the toughness of the interface and the stiffness of the material is a key
factor in choosing the TSL shape. DCB modelling and testing led to almost the same
results out of the four considered forms of the TSLs. However, when considering
the computational cost, the authors reported that the trapezoidal model is the worst
performing, since instability in the numerical analysis and convergence issues are
found in this model.

    Heidari-Rarani and Ghasemi [11] analyzed different CZM shapes, including trian-
gular, linear-exponential, trapezoidal, and trilinear for shear delamination propagation analysis in composite materials. The trilinear law used in their study was
constructed by superimposing two bilinear TSLs. For the material and geometry
tested and considering the investigated TSL shapes, the authors found that the bilinear
and linear-exponential shapes are not able to properly capture the behavior of the
tested specimens. However, the proposed trapezoidal and trilinear forms of CZM
were able to provide an accurate prediction. Although custom CZMs are usually
not simple to implement in commercial FE software, the authors affirm that their
proposed trilinear CZM form can be easily implemented in FE programs. Using
single lap adhesive joints with different types of adhesives, Campilho et al. [12]
evaluated the influence of TSLs on the performance of numerical models. To take
into account the effects of stress gradient on the results, joints with different overlap
geometries were selected. The authors found that for the more ductile adhesive used,
the trapezoidal shape gives a better estimate, while the bilinear and exponential TSLs
used for short overlaps underestimate and overestimate the mechanical response of
the joint, respectively. For the brittle adhesive, they found that all the considered
TSLs give comparable results although the bilinear shape CZM is the more precise.
They also found that the smaller the overlap length and the greater the ductility of
the adhesive the more influence the CZM shape has on the results. The shape of the
CZM was also analyzed by Shen and Chandra [8]. The concluded that the shape of
the CZM must represent the physical behavior of the material and attempted to find
a relationship between the shape of the CZM and the dissipation of thermomechan-
ical energy, the size of the plastic zone, the stiffness of the material, and the load at
which the crack starts. The authors report that the trapezoidal shape can take into

Fig. 2.6 A combination of
bilinear and trapezoidal
shape

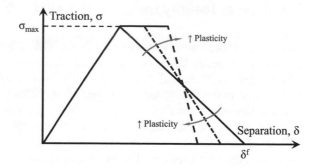

account the plastic deformation of the tested material, while the results of the other CZM shapes are not that precise. Considering the micro failure mechanisms, they claimed that the initial raising part of the TSL is related to some micro mechanisms such as the initiation of micro cracks and void growth and coalescence. Combining the bilinear and trapezoidal shapes, May et al. [18] proposed a modified TSL to take into account the effects of strain rate in adhesive joints with metallic substrates, reporting that the most suitable TSL shape also depends on the loading mode. Based on their results, for the tested material, a bilinear shape CZM can be employed for pure mode I but for pure mode II loading conditions the trapezoidal shape is better suited. However, the transition between the two shapes for mixed mode I + II conditions is a challenge when different CZM shapes are used for pure modes I and II. Accordingly, the authors proposed a pseudo-plasticity parameter able to solve this issue. The combined TSL shapes proposed by May et al. [18] are shown in Fig. 2.6.

By performing a numerical analysis of adhesive bonded composite materials in a scarf joint, Ridha et al. [19] found that using bilinear shape CZM the results are under-predicted. They also found that the exponential model is quite sensitive to the adhesive's toughness and not the other mechanical properties of the adhesive. However, their results showed that, for the tested configurations, the linear and trapezoidal forms of the CZM are significantly sensitive to the defined values of the maximum strength and fracture energy of the adhesive. According to their results using the exponential method, only the fracture energy of the adhesive must be precisely defined in the numerical analysis.

Overall, the literature suggests that for stable crack growth conditions, the shape of the TSL plays a key role in the accuracy of the obtained numerical results. Trapezoidal laws are often a highly suitable shape for these conditions. On the other hand, for more brittle adhesives in which the process zone in front of the crack tip is small enough (compared to the joint size) similar results will be obtained out of different TSL shapes. In this case using the simplest form of CZM which the bilinear shape is recommended.

2.3 How to Identify the CZM Parameters?

Achieving an accurate estimate in a numerical CZM analysis requires the TSL parameters of the problem to be defined correctly. Different approaches can be used to obtain TSL parameters. If the TSL is already known, these methods can be classified into two groups. In the first group, the shape of the CZM must be known in advance, while in the second group the shape traction–separation curve is formed dynamically during the analysis. One of the most common techniques used for identifying TSL parameters is the use of standard/routine strength and fracture testing. From these tests, the most important TSL parameters are obtained via direct measurement. But, as mentioned above, in this approach, the form of the TSL must be known in advance. Another commonly used technique for determining the CZM parameters is the inverse method. In this case, with the experimental results of a specific specimen geometry in hand, the parameters are calibrated until a good fit between the experimental and numerical results is reached. Again, the shape of the TSL must be determined in advance. Recently, a more powerful technique has been proposed, allowing not only the TSL parameters to be determined, but also the real CZM shape to be obtained directly from the tests. Accordingly, this approach is called the direct method. These methods are briefly discussed in the following sections.

2.3.1 Classic Characterization Method

The classic method is an approach in which the shape of the CZM should be determined or defined beforehand. Based on this method, standard/routine tests need to be conducted to calibrate the considered TSL. Although the bilinear shape is the most common TSL used with this methodology, other forms of CZM, such as the linear-exponential or the trapezoidal shapes are suitable for use in this approach. Modes I and II fracture energies, maximum traction in mode I, and the maximum strength in shear direction are the most important parameters that must be extracted from the characterization experiments. For the fracture energy, double cantilever beam (DCB) and end-notched flexure (ENF) specimens are commonly used to obtain the fracture energies for pure modes I and II, respectively. To this end, the raw load–displacement data obtained from the test should be treated using a suitable data reduction method. Different types of data reduction methods can be used, but those based on sample compliance generally lead to a more accurate result. While for composite materials the identification of the normal cohesive strength is still an open topic of research and different methods have been introduced by authors [20], for adhesive materials the maximum tensile traction is usually extracted from a tensile test conducted on a bulk dogbone sample of the cured adhesive. The Young's modulus of the adhesives is also calculated out of these tensile tests. Using the Young's modulus value, it is also possible to set the initial stiffness for a cohesive layer of a finite thickness. For shear conditions, equivalent properties can be obtained using the TAST (thick adherend shear test) methodology. It should be noted that the values of the initial stiffness for modes I and II depend on the thickness of the cohesive layer in the numerical analysis. This stiffness can sometimes be defined without any experiments (only for interfacial cohesive layers) and is thus called a penalty parameter, much higher than the actual stiffness of the material.

Despite the extensive use of the classic method to identify CZM parameters, this approach has some drawbacks. As mentioned above, the form of the TSL in this method must be defined prior to calibration tests and CZM calibration requires several tests performed on different types of joints and bulk samples. Although for DCB specimens the test has been standardized, the ENF tests have not yet been fully standardized.

2.3.2 Inverse Method

Paralleling the classic approach, in the inverse method the shape of the TSL should be defined initially. Although the material properties provide some insight regarding the proper form of the CZM, defining the form of the TSL in advance has the potential to introduce errors, which is seen as a disadvantage of the inverse approach. Nonetheless, the inverse method is still more precise than the classic method discussed above.

Conventional fracture tests can be employed for the inverse approach as well, although in this method no data reduction approach and post processing of the raw data are needed. By following a trial-and-error procedure, the CZM parameters should be adjusted until the best fit between the numerical results and the experimental load–displacement curve is obtained. The trial-and-error method can be part of a fully manual procedure, or it can be automated using an optimization algorithm. It should be noted that for each CZM analysis in pure mode loading conditions three different parameters should be set, however, only a single load–displacement curve out of the test is used to set these three parameters. Accordingly, the role of each parameter on the results should be well known to facilitate this adjustment process and to obtain the best fit with the experimental data.

2.3.3 Direct Method

Different parameters, such as stress triaxiality and joint geometry [21–23] influence the traction separation response of the adhesive. Despite the advantages of the classic and the inverse method, they are still not precise enough since the shape of the TSL is estimated before test. To overcome this issue, the direct method was developed where the shape of the TSL is precisely defined out of the data measured during the fracture characterization tests. To achieve this, the directly measured energy should be differentiated as a function of the measured displacement (separation). Using the direct method, both the shape of the TSL and the CZM parameters are obtained simultaneously. Different approaches have been proposed to analyze the behaviour of the specimens during the test. Measuring the rotation of the ends of the sample and the corresponding force during the test, using digital image correlation technique to analyze the displacement around the crack tip, and employing back face strain gauge to analyze the displacement of the adherends are the most commonly used techniques in the direct TSL determination approach. Ji et al. [24] and Zhu et al. [25] measured the relative rotation between the adherends at the crack tip. Considering an equivalent crack length technique Fernandes et al. [26] obtained cohesive law for pure mode II loading conditions using ENF samples. Using a DIC technique in a single edge notched beam (SENB) specimen tested under four-point bend loading conditions, Shen and Paulino [27] measured the full field displacements and this information was considered as the input for FE in a CZM analysis to obtain the TSL parameters. Later an analytical method was proposed based on the J-integral theory by Cui et al. [28] for mixed mode loading conditions. The tension and shear deformation was measured and analyzed using an in situ SEM (scanning electron microscope). More recently, Sun and Blackman [29] proposed a DIC method for obtaining J-integral, fracture energy and the shape of the TSL. They measured crack tip separation, beam rotation and crack length by DIC and used different data reduction approaches to obtain strain energy and TSL.

2.4   CZM Constitutive Laws

To implement CZM models into FE software packages it is first necessary to know how the traction and separation are related to each other. In this section the most widely employed constitutive laws of CZMs are listed.

2.4.1   Pure Mode Conditions

- Hillerborg model

As discussed before, one of the first and simplest CZM shapes was proposed by Hillerborg [4]. Due to the linear concept followed by this approach, it is mainly used for dealing with more brittle materials. According to the Hillerborg model, the relation between the traction and separation is defined as follows:

$$\sigma = \sigma_{\max}\left(1 - \frac{\delta}{\delta_0}\right) \tag{2.1}$$

where δ is the current value of the separation and δ_0 is the separation at the damage initiation point.

- **Cubic model**

The cubic TSL model proposed by Needleman [30] is similar to the CZM shape of Tvergaard [31] considering a cubic relation between the cohesive traction and separation. Due to its cubic behaviour, the Needleman model is more appropriate for ductile materials. The cubic TSL for the normal direction is defined as follows:

$$\sigma = -\frac{27}{4}\sigma_{max}\frac{\delta_I}{\delta_I^0}(1 - D_{max})^2 \qquad (2.2)$$

where D is defined as follows:

$$D = \sqrt{\left(\frac{\delta_I}{\delta_I^0}\right)^2 + \left(\frac{\delta_{II}}{\delta_{II}^0}\right)^2} \qquad (2.3)$$

The subscripts I and II denote the separation along the normal and shear direction respectively.

For pure mode I the relation is simplified as follows:

$$\sigma = -\frac{27}{4}\sigma_{max}\frac{\delta_I}{\delta_I^0}\left(1 - 2\frac{\delta}{\delta_I^0} + \left(\frac{\delta}{\delta_I^0}\right)^2\right) \qquad (2.4)$$

- **Bilinear relation**

The bilinear or triangular CZM shape is the most common TSL, used extensively in literature. This approach is already included in FE software packages and requires limited effort for its numerical implementation. Accordingly, due to the extensive application of the triangular CZM, this shape is discussed in more detail in this section.

A mode I triangular shaped CZM is shown in Fig. 2.7. Based on the bilinear TSL, the cohesive behaviour of the material is linear elastic until the maximum traction (σ_{max}). Until this point the cohesive damage is zero. By increasing the displacement damage increases until it completes when the separation reaches δ^f. The colored

Fig. 2.7 A typical triangular shaped CZM for tensile loading

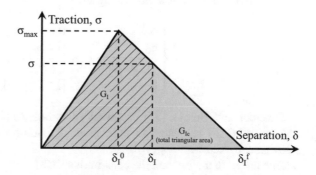

area shown in Fig. 2.7 represents the strain energy (G_I) corresponding to the current state of the separation (δ_I).

Accordingly, the constitutive law for the bilinear TSL is defined as follows:

$$\sigma = K[\delta_I - d(\delta_I + \langle -\delta_I \rangle)] \tag{2.5}$$

where σ is the tensile stress, $\langle\,\rangle$ is the McCauley bracket, and δ_I is the current normal separation. d in is the damage parameter defined as follows:

$$d = \frac{\delta^{\mathrm{f}}(|\delta_I| - \delta^0)}{|\delta_I|(\delta^{\mathrm{f}} - \delta^0)} \tag{2.6}$$

where $|\delta_I|$ is the absolute value of tensile separation. Note that compressive loads do not introduce any damage into the cohesive elements. To define the point where the damage initiates, several models have been proposed. One of them is the Maximum nominal stress criterion. In this approach the damage initiates whenever the maximum nominal stress (t) ratio reaches one as shown in Eq. 2.3.

$$\max\left\{\frac{\langle t_n \rangle}{t_n^{\max}} \cdot \frac{t_s}{t_s^{\max}} \cdot \frac{\langle t_t \rangle}{t_t^{\max}}\right\} = 1 \tag{2.7}$$

where the $t^{\max}$ represent the maximum traction and indices n, s, and t denote the normal, in plane shear, and out of plane shear directions, respectively.

A strain based approached can be also employed where a relationship similar to the stress-based model is defined (see Eq. 2.4).

$$\max\left\{\frac{\langle \varepsilon_n \rangle}{\varepsilon_n^0} \cdot \frac{\varepsilon_s}{\varepsilon_s^0} \cdot \frac{\varepsilon_s}{\varepsilon_t^0}\right\} \tag{2.8}$$

where ε^0 denotes the strain value at the damage initiation point.

- Trapezoidal model

Tvergaard and Hutchinson [14] later improved the model by proposing a new trilinear TSL where the traction is defined as follows:

$$\sigma = \begin{cases} K\delta & [if\ \delta \leq \delta_0] \\ \sigma_0 & [if\ \delta_0 \leq \delta \leq \delta^{pl}] \\ \frac{\sigma_0}{(\delta_f - \delta^{pl})}(\delta_f - \delta) & [if\ \delta^{pl} \leq \delta \leq \delta_f] \\ 0 & [if\ \delta \geq \delta_f] \end{cases} \tag{2.9}$$

Scheider and Brocks [32] proposed constitutive laws for a more generalized trapezoidal CZM subjected to pure mode I loading conditions. Their model is schematically shown in Fig. 2.8.

According to their model the constitutive CZM laws are defined as follows:

Fig. 2.8 A generalized trapezoidal TSL for pure mode I loading conditions

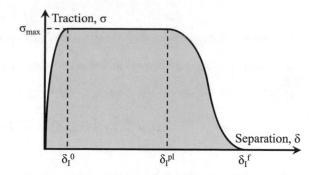

$$\sigma = \begin{cases} \sigma_0\left[2\frac{\delta}{\delta_0} - \left(\frac{\delta}{\delta_0}\right)^2\right] & [if\ \delta \leq \delta_0] \\ \sigma_0 & [if\ \delta_0 \leq \delta \leq \delta^{pl}] \\ 2\sigma_0\left(\frac{\delta-\delta^{pl}}{\delta_f-\delta^{pl}}\right)^3 - 3\left(\frac{\delta-\delta^{pl}}{\delta_f-\delta^{pl}}\right)^2 + 1 & [if\ \delta^{pl} \leq \delta \leq \delta_f] \\ 0 & [if\ \delta \geq \delta_f] \end{cases} \tag{2.10}$$

The symbols used in these relations are defined in Fig. 2.8.

- Bilinear softening model

A bilinear softening approach was proposed by Bazant [33]. Based on the Bazant approach the TSL constitutive laws are defined as follows:

$$\sigma = \begin{cases} \sigma_{max}\left(1 - \frac{\delta}{\delta_1} + \frac{\sigma_1\delta}{\sigma_{max}\delta_1}\right) & [if\ \delta \leq \delta_1] \\ \sigma_I\left(1 - \frac{\delta-\delta_1}{\delta_f-\delta_1} + \frac{\sigma_1\delta}{\sigma_{max}\delta_1}\right) & [if\ \delta_1 \leq \delta \leq \delta_f] \end{cases} \tag{2.11}$$

- Linear-parabolic

A linear parabolic curve was considered by some authors such as [5] to simulate the interfacial failure in mode I and mode II loading conditions. In the linear-parabolic model the TSL is defined as follows:

$$\sigma = \begin{cases} K\delta & [if\ \delta \leq b_0] \\ max\left(0.\frac{\sigma_0}{2} + \frac{\sigma_0}{\delta_f-\delta_0}(\delta - b_0) - \frac{\sigma_0}{\delta_f-\delta_0^2}(\delta - \delta_0)^2\right) & [if\ b_0 \leq \delta \leq \delta_f] \end{cases}$$

$$b_0 = \frac{\sigma_0}{2K} \tag{2.12}$$

where b_0 shows the point where the traction reaches its maximum.

- Exponential law

The exponential law is already implemented in some FE software packages such as Abaqus. The constitutive laws for an exponential CZM shape are as follows:

$$\sigma = \begin{cases} K\delta e^{-\frac{\delta}{c_0}} & [if \ \delta \leq c_0] \\ \sigma_0[1 + \beta(\delta - c_0)]e^{-\beta(\delta-c_0)} & [if \ c_0 \leq \delta \leq \delta_f] \end{cases} \tag{2.13}$$

where β must be obtained through an iterative approach using the following relation [5]:

$$D = D(\beta) \int_0^\infty \sigma(\delta.\beta) d\delta = G_c \tag{2.14}$$

and C_0 is the separation at which the traction reaches its maximum value (σ_0) and can be obtained through the following relation.

$$c_0 = \frac{\sigma_0 e}{K} \tag{2.15}$$

Needleman [34] also proposed an exponential model suitable for mode I loading conditions where the traction separation was defined as follows:

$$\sigma = -\frac{16}{9}\sigma_{max}e^2\frac{\delta_I}{\delta_{I0}}\exp\left(-\frac{16}{9}e\frac{\delta_I}{\delta_{I0}}\right) \tag{2.16}$$

- Modes II and III laws

Mode II and III constitutive laws are similar to tensile TSLs. However, as discussed above for mode I, the compressive load does not introduce any damage into the cohesive elements while for the shear mode both the positive and negative shear components damage the bondline (see Fig. 2.9). The hatched area shown in Fig. 2.9 represents the shear strain energy (GII) based on the current state of traction and separation.

On the other hand, since mode II and mode III are both shear conditions, in 3D models an equivalent shear is defined using both the in plane and out of plane shear stress and separation components. Accordingly, the equivalent shear deformation is defined as follows:

$$|\delta_{shear}| = \sqrt{\delta_{II}^2 + \delta_{III}^2} \tag{2.17}$$

where δ_{II} and δ_{III} correspond to the in plane and out of plane shear separation, respectively. Accordingly, the separation at which the damage initiates is defined as follows:

$$\delta_{shear}^0 = \left[\left(\delta_{II}^0\right)^2 + \left(\delta_{III}^0\right)^2\right]^{1/2} \tag{2.18}$$

And consequently the maximum equivalent shear separation is obtained by the following relation:

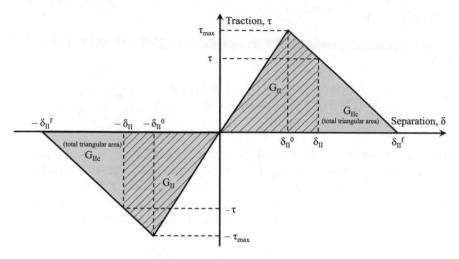

Fig. 2.9 Typical triangular shaped CZM for shear loading condition

$$\delta_{shear}^{f} = \left[\left(\delta_{II}^{f} \right)^2 + \left(\delta_{III}^{f} \right)^2 \right]^{1/2} \tag{2.19}$$

Using the equivalent shear strain, the cohesive damage for shear loading condition can be obtained by the following equation.

$$d = \frac{\delta_{shear}^{f} \left(|\delta_{shear}| - \delta_{shear}^{0} \right)}{|\delta_{shear}| \left(\delta_{shear}^{f} - \delta_{shear}^{0} \right)} \tag{2.20}$$

By knowing the damage value, the constitutive relation between the shear traction and shear separation is defined as follows:

$$\tau = K_0 [\delta_{shear} - d\delta_{sheasr}] \tag{2.21}$$

2.4.2 Mixed Mode Constitutive Laws

Considering the practical applications where the stress state in bonded joints is of mixed nature, mixed mode cohesive zone models have received significant attention in recent years [18, 35–40]. The most commonly used techniques considered in literature to mix the pure mode TSLs are discussed in this section.

- Bilinear models

Cohesive traction and strain for different parts of the mixed mode are related to each other as follows:

$$\begin{Bmatrix} \sigma_n \\ \tau_{sn} \\ \tau_{tn} \end{Bmatrix} = (1-d)K \begin{Bmatrix} \varepsilon_n \\ \varepsilon_{sn} \\ \varepsilon_{tn} \end{Bmatrix} - dK \begin{Bmatrix} \langle -\varepsilon_n \rangle \\ 0 \\ 0 \end{Bmatrix} \quad \langle x \rangle = \begin{cases} x; \ x > 0 \\ 0; \ x \leq 0 \end{cases} \tag{2.22}$$

where, σ_n, τ_{sn} and τ_{tn} are the normal, in plane shear and out of plane shear stresses in a cohesive element, respectively. Figure 2.10 schematically shows a 3D cohesive element in which the above-mentioned parameters are shown. K in Eq. 2.22 is the initial stiffness of the elements. Strain ε in cohesive elements is related to the separation δ by the following relations in which the h_0 shows the thickness of the cohesive elements.

$$\varepsilon_n = \frac{\delta_n}{h_0}; \quad \varepsilon_{sn} = \frac{\delta_{sn}}{h_0}; \quad \varepsilon_{tn} = \frac{\delta_{tn}}{h_0} \tag{2.23}$$

where the subscripts n, sn and tn denote the normal, in-plane shear and out-of-plane shear modes, respectively.

The damage parameter (d) in Eq. 2.22 is defined as follows:

$$d = \frac{\varepsilon_m^f \left(\varepsilon_m - \varepsilon_m^0 \right)}{\varepsilon_m \left(\varepsilon_m^f - \varepsilon_m^0 \right)} \tag{2.24}$$

where for 2D conditions the equivalent strain ε_m is defined as follows

$$\varepsilon_m = \sqrt{\langle \varepsilon_n \rangle^2 + \varepsilon_{sn}^2} \tag{2.25}$$

where ε_m^0 is the equivalent damage initiation strain and ε_m^f is the maximum equivalent strain. Both the damage initiation strain and the damage evolution strain are obtained

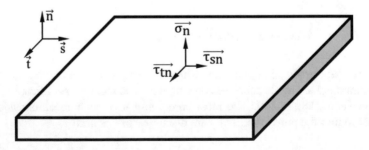

Fig. 2.10 3D stress state in an interface element

as follows [41], where for the ε_m^f the BK (Benzeggagh–Kenane) method is used.

$$\varepsilon_m^0 = \begin{cases} \varepsilon_{sn}^0 \varepsilon_n^0 \sqrt{\dfrac{1+\beta^2}{(\varepsilon_{sn}^0)^2 + (\beta \varepsilon_n^0)^2}} & [if \; \sigma_n > 0] \\ \varepsilon_{sn}^0 & [if \; \sigma_n \leq 0] \end{cases} \tag{2.26}$$

where:

$$\varepsilon_{sn}^0 = \frac{\tau_{sn}^0}{K}; \quad \varepsilon_n^0 = \frac{\sigma_n^0}{K} \tag{2.27}$$

$$\beta = \frac{\varepsilon_{sn}}{\varepsilon_n} \tag{2.28}$$

$$\varepsilon_m^f = \begin{cases} \dfrac{2}{h_0 K \varepsilon_m^0} \left[G_{Ic} + (G_{IIc} - G_{Ic}) \left\{ \dfrac{\beta^2}{1+\beta^2} \right\}^\eta \right] & [if \; \sigma_n > 0] \\ \varepsilon_{ns}^f & [if \; \sigma_n \leq 0] \end{cases} \tag{2.29}$$

Figure 2.11 schematically shows the concepts of mixed mode in a bilinear TSL.

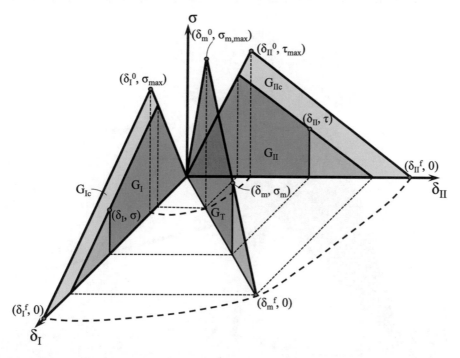

Fig. 2.11 Mixed mode concepts in bilinear TSL

- Coupled trapezoidal and bilinear models

As discussed before, in some studies [18] the effect of mixed bilinear-trapezoidal CZM was considered where the bilinear shape was used for mode I condition while shear mode was simulated using a trapezoidal shape CZM. Due to the difference in the shapes of the CZM, constitutive models for mixed mode loading are different from those discussed in the previous section. Figure 2.12 shows the scheme of a coupled bilinear and trapezoidal (trilinear) TSLs.

May et al. [18] used the coupled triangle-trapezoidal method to simulate the strain rate effects in different bonded joints. According to their model, the following constitutive laws are defined for the introduced couple cohesive model.

For the point where the damage initiation starts, an exponential law was defined which is common in bilinear shape models.

$$\left(\frac{\sigma_I}{\sigma_{I0}}\right)^{\alpha} + \left(\frac{\sigma_{II}}{\sigma_{II0}}\right)^{\alpha} = 1 \tag{2.30}$$

where σ_I and σ_{II}, are the stress components correspond to the normal and shear directions and σ_I^0, and σ_{II}^0 are the maximum mode I and mode II tractions where the damage initiation starts. α is a fitting parameter that is usually set in the range of 1–2.

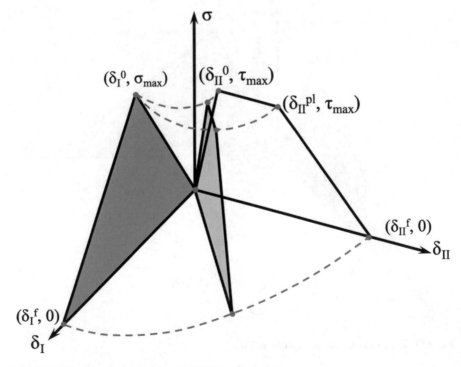

Fig. 2.12 Combined trapezoidal and triangle CZM shapes

Using the BK method, the mixed mode energy is defined as follows:

$$G_c = G_{Ic} + (G_{IIc} - G_{Ic})\left(\frac{\beta^2 K_{II}}{K_I + \beta^2 K_{II}}\right)^\eta \tag{2.31}$$

where the mode ratio is defined based on the separation components as follows:

$$\beta = \frac{\delta_I}{\delta_m} \tag{2.32}$$

The damage initiation parameters and the maximum traction at which the complete failure of the element occurs can be defined as follows [18]:

$$\delta_m^0 = \delta_I^0 \delta_{II}^0 \sqrt{\frac{1 + \beta^2}{\delta_{II}^0{}^2 + \left(\beta \delta_I^0\right)^2}} \tag{2.33}$$

$$\sigma_m^0 = \frac{\delta_m^0}{\sqrt{1 + \beta^2}} \sqrt{K_I^2 + (\beta K_{II})^2} \tag{2.34}$$

$$\delta_m^f = \frac{2G_c}{\sigma_m^0} + \delta_m^0 - \delta_m^{pl} \tag{2.35}$$

$$\delta_m^{pl} = \delta_I^0 \delta_{II}^{pl} \sqrt{\frac{1 + \beta^2}{\left(\delta_{II}^{pl}\right)^2 + \left(\beta \delta_I^0\right)^2}} \tag{2.36}$$

$\delta_I^P, \delta_{II}^P$, and δ_m^P are the tensile separation, shear separation, and the equivalent separation for the mode ratio of β, respectively.

As shown above, one of the key parameters in mixed mode constitutive CZM laws is the mode ratio. There are different approaches to define the mode ratio. Mode mixity can be defined as a function of energy, traction, or separation. The mode ratio obtained out of each method can be different. When a joint is subjected to a pure mode I conditions are defined by energy components, the shear energy is zero while a non-zero value is assigned to the tensile energy. However, in this condition both the tensile and shear tractions can be non-zero. This happens due to the coupled traction separation behaviour that occurs for some materials. Consequently, despite the pure mode I state based on energy values, the traction-based mode ratio does not show pure mode conditions.

2.5 Fatigue Analysis Using CZM

The CZM was considered not only for static loading conditions, but also for assessing the fatigue life of bonded joints. Different aspects must be considered for using CZM for the fatigue analysis of adhesives. First, the general fatigue behavior of the adhesive must be known. Different parameters such as adhesive type, joint geometry, humidity, ambient temperature, and also loading conditions (R ratio, frequency, etc.) affect the fatigue behavior of an adhesive joint. Consequently, the adhesive can experience high cycle fatigue with a significant fatigue damage initiation life or a short fatigue life, where significant plasticity occurs within the adhesive layer during cyclic loading. These aspects were discussed in Chap. 1. The next step is to choose the most suitable CZM form. Each CZM has its own specificities and based on the fatigue behavior of the joint, a suitable CZM should be selected. In this chapter, the concepts of CZM and different TSLs were discussed and the corresponding constitutive laws were presented. When choosing the CZM law, the next step is to modify it to consider the effect of fatigue loading on the degradation of the adhesive's cohesive properties. This aspect is discussed in Chap. 3, where fatigue simulation of adhesive joints using cohesive modeling is explained and different CZM-based fatigue models proposed in the literature are reviewed.

References

1. Dugdale, D.S. 1960. Yielding of steel sheets containing slits. *Journal of the Mechanics and Physics of Solids* 8 (2): 100–104.
2. Barenblatt, G.I. 1962. The mathematical theory of equilibrium cracks in brittle fracture. In *Advances in Applied Mechanics*, 55–129. Elsevier.
3. Wciślik, W., and T. Pała. 2021. Selected aspects of cohesive zone modeling in fracture mechanics. *Metals* 11 (2): 302.
4. Hillerborg, A., M. Modéer, and P.-E. Petersson. 1976. Analysis of crack formation and crack growth in concrete by means of fracture mechanics and finite elements. *Cement and Concrete Research* 6 (6): 773–781.
5. Alfano, G. 2006. On the influence of the shape of the interface law on the application of cohesive-zone models. *Composites Science and Technology* 66 (6): 723–730.
6. Allix, O., and A. Corigliano. 1996. Modeling and simulation of crack propagation in mixed-modes interlaminar fracture specimens. *International Journal of Fracture* 77 (2): 111–140.
7. Chen, J. 2002. Predicting progressive delamination of stiffened fibre-composite panel and repaired sandwich panel by decohesion models. *Journal of Thermoplastic Composite Materials* 15 (5): 429–442.
8. Chandra, N., et al. 2002. Some issues in the application of cohesive zone models for metal–ceramic interfaces. *International Journal of Solids and Structures* 39 (10): 2827–2855.
9. Kafkalidis, M., and M. Thouless. 2002. The effects of geometry and material properties on the fracture of single lap-shear joints. *International Journal of Solids and Structures* 39 (17): 4367–4383.
10. Alfano, M., et al. 2009. Mode I fracture of adhesive joints using tailored cohesive zone models. *International Journal of Fracture* 157 (1–2): 193.
11. Heidari-Rarani, M., and A. Ghasemi. 2017. Appropriate shape of cohesive zone model for delamination propagation in ENF specimens with R-curve effects. *Theoretical and Applied Fracture Mechanics* 90: 174–181.
12. Campilho, R.D., et al. 2013. Modelling adhesive joints with cohesive zone models: Effect of the cohesive law shape of the adhesive layer. *International Journal of Adhesion and Adhesives* 44: 48–56.

13. Song, S.H., G.H. Paulino, and W.G. Buttlar. 2008. Influence of the cohesive zone model shape parameter on asphalt concrete fracture behavior. In *AIP Conference Proceedings*. American Institute of Physics.
14. Tvergaard, V., and J.W. Hutchinson. 1992. The relation between crack growth resistance and fracture process parameters in elastic-plastic solids. *Journal of the Mechanics and Physics of Solids* 40 (6): 1377–1397.
15. Xu, X.-P., and A. Needleman. 1993. Void nucleation by inclusion debonding in a crystal matrix. *Modelling and Simulation in Materials Science and Engineering* 1 (2): 111.
16. He, M.-H., and K.-G. Xin. 2011. Separation work analysis of cohesive law and consistently coupled cohesive law. *Applied Mathematics and Mechanics* 32 (11): 1437–1446.
17. Gao, Y., and A. Bower. 2004. A simple technique for avoiding convergence problems in finite element simulations of crack nucleation and growth on cohesive interfaces. *Modelling and Simulation in Materials Science and Engineering* 12 (3): 453.
18. May, M., H. Voß, and S. Hiermaier. 2014. Predictive modeling of damage and failure in adhesively bonded metallic joints using cohesive interface elements. *International Journal of Adhesion and Adhesives* 49: 7–17.
19. Ridha, M., V. Tan, and T. Tay. 2011. Traction–separation laws for progressive failure of bonded scarf repair of composite panel. *Composite Structures* 93 (4): 1239–1245.
20. Moslemi, M., and M. Khoshravan. 2015. Cohesive zone parameters selection for mode-I prediction of interfacial delamination. *Strojniski Vestnik/Journal of Mechanical Engineering* 61 (9).
21. Akhavan-Safar, A., et al. 2018. The role of T-stress and stress triaxiality combined with the geometry on tensile fracture energy of brittle adhesives. *International Journal of Adhesion and Adhesives* 87: 12–21.
22. Akhavan-Safar, A., et al. 2021. Fracture energy assessment of adhesives–Part I: Is GIC an adhesive property? A neural network analysis. *Proceedings of the Institution of Mechanical Engineers, Part L: Journal of Materials: Design and Applications* 235 (6): 1461–1476.
23. Delzendehrooy, F., et al. 2021. Fracture energy assessment of adhesives part II: Is GIIc an adhesive material property? (A neural network analysis). *Journal of Advanced Joining Processes* 3: 100049.
24. Ji, G., et al. 2010. Effects of adhesive thickness on global and local mode-I interfacial fracture of bonded joints. *International Journal of Solids and Structures* 47 (18–19): 2445–2458.
25. Zhu, Y., K.M. Liechti, and K. Ravi-Chandar. 2009. Direct extraction of rate-dependent traction–separation laws for polyurea/steel interfaces. *International Journal of Solids and Structures* 46 (1): 31–51.
26. Fernandes, P., et al. 2017. The influence of water on the fracture envelope of an adhesive joint. *Theoretical and Applied Fracture Mechanics* 89: 1–15.
27. Shen, B., and G. Paulino. 2011. Direct extraction of cohesive fracture properties from digital image correlation: A hybrid inverse technique. *Experimental Mechanics* 51 (2): 143–163.
28. Cui, W., and M. Wisnom. 1993. A combined stress-based and fracture-mechanics-based model for predicting delamination in composites. *Composites* 24 (6): 467–474.
29. Sun, F., and B. Blackman. 2020. A DIC method to determine the Mode I energy release rate G, the J-integral and the traction-separation law simultaneously for adhesive joints. *Engineering Fracture Mechanics* 234: 107097.
30. Needleman, A. 1987. A continuum model for void nucleation by inclusion debonding.
31. Tvergaard, V. 1989. Material failure by void growth to coalescence. *Advances in Applied Mechanics* 27: 83–151.
32. Scheider, I., and W. Brocks. 2003. Simulation of cup–cone fracture using the cohesive model. *Engineering Fracture Mechanics* 70 (14): 1943–1961.
33. Bazant, Z. 1993. *38 Current Status and Advances in the Theory of Creep and Interaction With Fracture.*
34. Needleman, A. 1990. An analysis of decohesion along an imperfect interface. In *Non-linear Fracture*, 21–40. Springer.

35. de Oliveira, L.A., and M.V. Donadon. 2020. Delamination analysis using cohesive zone model: A discussion on traction-separation law and mixed-mode criteria. *Engineering Fracture Mechanics* 228: 106922.
36. Sun, L., et al. 2020. Prediction of failure behavior of adhesively bonded CFRP scarf joints using a cohesive zone model. *Engineering Fracture Mechanics* 228: 106897.
37. Rocha, A., et al. 2020. Numerical analysis of mixed-mode fatigue crack growth of adhesive joints using CZM. *Theoretical and Applied Fracture Mechanics* 102493.
38. Lapczyk, I., and J.A. Hurtado. 2007. Progressive damage modeling in fiber-reinforced materials. *Composites Part A: Applied Science and Manufacturing* 38 (11): 2333–2341.
39. Tie, Y., et al. 2018. An insight into the low-velocity impact behavior of patch-repaired CFRP laminates using numerical and experimental approaches. *Composite Structures* 190: 179–188.
40. Hou, Y., et al. 2019. Low-velocity impact behaviors of repaired CFRP laminates: Effect of impact location and external patch configurations. *Composites Part B: Engineering* 163: 669–680.
41. Ye, L. 1988. Role of matrix resin in delamination onset and growth in composite laminates. *Composites Science and Technology* 33 (4): 257–277.

Chapter 3
Fatigue Degradation Models

Abstract Although the selection of an appropriate cohesive zone modelling (CZM) shape and the identification of the cohesive parameters are important parts of the fatigue life analysis of bonded joints, it is still necessary to understand how cohesive properties vary (degrade) with the load cycles. This chapter addresses this point, where different CZM based fatigue life prediction models are reviewed and discussed. In this chapter, cohesive based fatigue methods are classified into two main groups, including the loading envelope (cycle jumping) strategy approaches and the cycle-by-cycle analysis models. The first one is mainly used for high cycle fatigue regimes and the second one is more suitable for low cycle fatigue where the fatigue stress level is close or above the yielding point of the adhesive. According to the cycle jumping model, a fatigue loading can be simulated by a static load. Two techniques called LDFA and EXFIT discussed in this chapter follow the load envelope strategy. In LDFA a link is created between the damage mechanics and fracture mechanics. Several models based on the LDFA have been developed by proposing new damage mechanics or by modifying the fracture mechanics part. Another load envelope technique discussed in this chapter is EXFIT where the fatigue damage parameter is defined as a function of an effective stress or strain parameter. In terms of the numerical implementation, the models based on the LDFA need further work than the EXFIT approaches. In addition to these techniques, a cycle-by-cycle method was also considered in some studies. Simulating actual fatigue cycles rather than using a constant static load is the main difference between load envelope strategies and the cycle-by-cycle approach. Cycle-by-cycle fatigue analysis is based on the concepts of irreversible CZMs, where cohesive stiffness in reloading is not the same as in unloading. All of these techniques and their relationships are discussed in this chapter.

3.1 Introduction

The first step in fatigue life analysis of adhesive joints using CZM is the definition of the initial TSL. Different TSLs were introduced and discussed in Chap. 2. In the current chapter, we will discuss how the initial shape of the TSL should be modified

© The Author(s), under exclusive license to Springer Nature Switzerland AG 2022   43
A. Akhavan-Safar et al., *Cohesive Zone Modelling for Fatigue Life Analysis of Adhesive Joints*, SpringerBriefs in Computational Mechanics,
https://doi.org/10.1007/978-3-030-93142-1_3

to be able to consider a degradation of cohesive properties as a function of load cycles. To adapt the initial shape of the CZM for fatigue analysis, additional aspects should be defined for the TSL. The first is considering the loading–unloading (reloading-unloading) path. Since the adhesive subjected to cyclic loads will experience both loading and unloading, the unloading and reloading path should be defined for the considered TSL. As mentioned above, the cohesive elements subjected to fatigue loading experience two different damage types. The first is the static damage caused by the static load applied to the elements. This damage is already discussed in Chap. 2. However, due to the cyclic loads, the cohesive elements undergo another source of damage caused by the cyclic load called fatigue damage. Accordingly, the TSL which is defined for fatigue loading conditions should be able to take into account both the static and the fatigue damages. To this end, the defined TSL should be reshaped, according to the number of loading cycles that the cohesive element has experienced. Although not fundamental for all loading conditions and joint geometries, the definition of the frictional behavior of the elements due to contact of the crack surface is another aspect that should be considered in fatigue analysis of bonded joints.

Out of the three aspects mentioned above, the first two are considered as more relevant by researchers in the field. There are two strategies to consider fatigue damage evolution in cohesive elements. The first strategy is based on the load-unload technique. According to this approach, the initial stiffness of the elements is degraded cycle by cycle. The rate of degradation is usually obtained experimentally. This method is called cycle-by-cycle analysis. In the cycle-by-cycle analysis, the elements should be subjected to the real fatigue loads that the joint experiences in practice. This approach is appropriate for low cycle fatigue regime where the joints fail usually in less than one thousand cycles.

For a high cycle fatigue regime, a cycle-by-cycle analysis will significantly increase the computational time and cost. The degradation of the stiffness or the cohesive properties is not significant per loading cycle and consequently, one can define a jumping factor to analyze, for example, the cohesive properties every n cycles. Here n is defined case by case and can be for example 100, or 1000 or even a higher value. This strategy is called cycle jumping or a load envelope approach. This approach is based on the linear elastic fracture mechanics assumptions and, as explained above, is good for high cycle fatigue conditions.

There are two methods to calculate the damage accumulation in cohesive elements using the cycle jumping (load envelope) strategy. The first technique is linking the damage mechanics concepts to the fracture mechanics as initially proposed by Turon et al. [1]. In this approach, the model parameters are directly calculated out of the fatigue crack growth tests conducted in pure mode I (using DCB specimens) and in pure mode II (using ENF samples). Another technique is the fitting approach. In this method, damage propagation is defined as a function of an equivalent stress [2] or strain [3–5] component. The rate of damage evolution is obtained by fitting the numerical results to the experimental data. Accordingly, a trial-and-error procedure is required to obtain the correct values of the model parameters, a process which is often seen as the main disadvantage of the fitting approach.

In the following sections both the cycle by cycle and load envelope (cycle jumping) methods are discussed in greater detail.

3.2 Loading Envelope Strategy

3.2.1 Linking Damage and Fracture Mechanics Approach (LDFA)

As mentioned above, one of the most important differences between the static and fatigue CZMs is the additional fatigue damage added to the static damage. The main question that should be answered here is what is the fatigue damage and how the rate of fatigue damage increases per loading cycle. Accordingly, the rate of damage growth as a function of load cycle (dD/dN) must be calculated and various methods have been proposed to this end. Linking damage mechanics and fracture mechanics is a solution that helps to determine the rate of fatigue damage, based on Eq. 3.1.

$$\frac{dD}{dN} = \frac{dD}{dA_d} \times \frac{dA_d}{dN} \tag{3.1}$$

where A_d is the damaged area in the cohesive zone and N is the cycle number.

In Eq. 3.1, dD/dA_d is calculated based on the damage mechanics concepts and the last term (dA_d/dN) should be obtained according to the LEFM approach.

For the *ith* increment the new cracked area $(\Delta A_d)_i$ is obtained by the summation of the cracked areas of the integration points (IPs) within the cohesive zone using the following relation:

$$(\Delta A_d)_i = \sum_{J=1}^{n_{cz}} A^J (\Delta d)_i^J = \Delta A_i \tag{3.2}$$

where n_{cz} denotes the number of IPs in the cohesive zone area (where $0 < d < 1$), A^J is the effective area of Jth integration point. As shown schematically in Fig. 3.1 where the equivalent cracked length is the summation of the physical crack size plus the crack length calculated using Eq. 3.2.

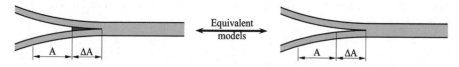

Fig. 3.1 Schematic representation of the equivalence between the increase of the damaged area and the crack growth

Considering Eq. 3.2, the new cracked (damaged) area within the process zone can be rewritten as follows:

$$\frac{\mathrm{d}A}{\mathrm{d}N} = \sum_{Acz} \frac{\mathrm{d}A_d^i}{\mathrm{d}N} \tag{3.3}$$

where A_{cz} is the area of the damaged zone in the adhesive layer. The left side of Eq. 3.3 is obtained using the LEFM concepts and Paris law relations. Accordingly, Eq. 3.3 relates the damage mechanics to the LEFM concepts. Several Paris law models have been developed and verified. The capabilities of some of the most common Paris law models for bonded structures were summarized and evaluated in [6]. The goal in LDFA is to separately define the two parts of Eq. 3.1 and then link them to finally obtain the rate of fatigue damage accumulation within the adhesive layer in the bonded joint.

LDFA was first proposed by Turon et al. [1]. According to their approach the fracture mechanics and damage mechanics terms in Eq. 3.1 were defined as follows:

$$\frac{\partial d}{\partial A_d} = \frac{1}{A^e} \frac{\left[\Delta^f (1 - d) + d\Delta^0\right]^2}{\Delta^0 \Delta^f} \tag{3.4}$$

$$\frac{\partial A_d}{\partial N} = \frac{A^e}{A_{CZ}} \frac{\partial A}{\partial N} \tag{3.5}$$

Considering Eqs. 3.4 and 3.5 the rate of damage evolution as a function of the fatigue cycles is obtained as follows:

$$\frac{\partial d}{\partial N} = \frac{1}{A_{CZ}} \frac{\left[\Delta^f (1 - d) + d\Delta^0\right]^2}{\Delta^0 \Delta^f} \frac{\partial A}{\partial N} \tag{3.6}$$

where (A_d) is the damaged area in a cohesive element and the effective area in this zone is shown by (A^e).

The LDFA has been significantly improved in recent years with the development of new methodologies for both the fracture mechanics and damage mechanics components. Drawing inspiration from [7], Moroni and Pirondi et al. [8] proposed a fatigue model and simulated the fatigue crack growth in bonded joints using Abaqus. Relating the crack growth rate with the damage evolution, they proposed a fatigue model for the pure modes I and II and the mixed modes I + II of fatigue loading. According to their model, the damage evolution rate as a function of the load cycle is defined as follows:

$$\frac{\mathrm{d}D}{\mathrm{d}N} = \frac{1}{A_{cz}} B \Delta G^d \tag{3.7}$$

In most of the LDFA based models the size of the cohesive zone should be calculated manually, which requires the interruption of the numerical analysis. However,

Moroni and Pirondi [8] proposed a fully automated procedure, where the fatigue crack growth within the adhesive layer is simulated in a single run.

In another study, de Moura and Gonçalves [9], introduced a cohesive zone method for pure shear loading conditions using the LDFA assumptions. They used a modified Paris law model. In their approach, the damage mechanics term of the LDFA model was defined as follows:

$$\frac{\mathrm{d}e_f}{\mathrm{d}A_{pk}} = \frac{u_u u_0}{u_{max}^2 A_{tk}} \tag{3.8}$$

where u_u is the ultimate displacement, u_0 is the separation at damage initiation and u_{max} is the maximum relative separation during the fatigue loading history. A_{tk} in Eq. 3.8 is the area of each integration point which depends on the mesh refinement. On the other hand, using the weight of each integration point (r_{wk}) the fracture mechanics term of the LDFA was defined as follows:

$$\frac{\mathrm{d}A_{pk}}{\mathrm{d}N} = \frac{C_1 r_{wk}}{n_{fpz}} \left(\frac{\Delta G_{IIk}}{G_{IIc}} \right)^{C_2} \tag{3.9}$$

where C_1 and C_2 are Paris law constants and should be obtained experimentally. N_{fpz} is the number of integration points in the cohesive zone.

Considering Eqs. 3.8 and 3.9, the rate of fatigue damage evolution can be calculated as follows:

$$\frac{\mathrm{d}e_f}{\mathrm{d}N} = \frac{u_u u_0}{u_{max}^2 A_{tk}} \frac{C_1 r_{wk}}{n_{fpz}} \left(\frac{\Delta G_{IIk}}{G_{IIc}} \right)^{C_2} \tag{3.10}$$

The same authors extended their model to mixed mode loading conditions [10] and evaluated it with mixed mode I + II fatigue analysis of adhesively bonded composite joints. To enable the use of LDFA for mixed mode conditions, an additional step is required compared to the pure mode conditions. One of the most important parts of mixed-mode LDFA models is knowing how to mix the fatigue results of pure modes I and II into a single Paris law equation. Section 3.3 discusses this point.

Pirondi et al. [11] proposed a fatigue crack growth model for 3D cracks and general shapes. They used a model similar to that already developed by Moroni and Pirondi [8] in which the two terms of the LDFA are defined as follows:

Damage mechanics term:

$$\frac{\mathrm{d}D}{\mathrm{d}A_d} = \frac{1}{A_e} \tag{3.11}$$

LEFM term:

$$\frac{\mathrm{d}A_d}{\mathrm{d}N} = \frac{\mathrm{d}A}{\mathrm{d}N} \frac{A_e}{A_{cz}} \tag{3.12}$$

The area of the cohesive zone (A_{CZ}) in these equations effectively starts from the crack tip up to the point at which the maximum traction is attained. Although the value of A_{cz} (called L_{cz} for 2D conditions) is often obtained numerically [12–14] by conducting a static analysis before the fatigue simulation, the length of the fracture process zone (A_{cz} or L_{cz}) for pure mode I loading conditions can be automatically calculated using the following relation, based on the relation of Rice [15]:

$$L_{CZ} = b \frac{9\pi}{32} \frac{E G_{max}}{\sigma_{max}^2} \tag{3.13}$$

where E is the Young's modulus of the substrates, s_{max} is the maximum traction, G_{max} is the maximum strain energy release rate at each loading cycle, and b is the width of the joint. The size of the A_{cz} was also analyzed by Harper and Hallett [16]. They proposed the following relations separately for pure mode I and pure mode II loading conditions:

$$A_{cz.I} = bE' \frac{G_{Ic}}{\sigma_0^2} \tag{3.14a}$$

$$A_{cz.II} = bE' \frac{G_{IIc}}{\tau_0^2} \tag{3.14b}$$

where b is the width of the joint and G_{Ic} and G_{IIc} are the tensile and shear fracture energies. E' in these equations is the Young's modulus for materials with isotropic properties and an equivalent elastic modulus for orthotropic materials. Naghipour et al. [17] extended the works of Rice [15] and Harper and Hallet [16] for mixed mode loading conditions. According to their approach the mixed mode cohesive zone size ($A_{cz,mixed\,mode}$) is defined as the minimum possible area of the mode I and mode II cohesive zones multiplied by a scaling factor (M) which mathematically is shown as follows:

$$A_{cz.mixed\,mode} = M(\text{minimum}(Eqs. \ 3.14a \ and \ 3.14b)) \tag{3.15}$$

The scaling factor M which is also considered as a calibration parameter should be set to get the best fit between the experimental and numerical analysis. Accordingly, the value of 0.5 was used in the Harper and Hallet model [16] and it was set to 0.65 in the work of Naghipour et al. [17].

As discussed above, some of the LDFA models employ an analytical approach to calculate the A_{cz} (L_{cz}). However, using these methods the effects of the static damage on the fatigue damage accumulation in the process zone is ignored. Unwanted fatigue damage (sometimes called static damage) stemming from the evolution of the damage caused by fatigue loading, can influence the fatigue life of the cohesive elements. Since in the load envelope strategy the applied load is kept constant during the analysis, the damage level will be constant as well and consequently the life is infinite. Therefore, to estimate the fatigue life, fatigue damage must be introduced to the cohesive layer, degrading the cohesive properties as a function of the load cycles. On the other hand, in the cycle jumping, technique the elements are subjected to a constant load, therefore any degradation in cohesive stiffness will result in an additional displacement and consequently will lead to a higher damage level called unwanted fatigue (static) damage.

This phenomenon has been considered in some studies [12, 18, 19]. Using the J-integral method at crack tip instead of using the data of the integration points and by considering different mode mixities and for both the constant and variable amplitudes Oliveira and Donadon [19] analyzed the effects of the unwanted static damage in the fatigue damage. Based on their approach the fatigue damage D_A^f is defined as:

$$D_A^f = \frac{a_{el}^f}{l_{el}} \left(1 - D_A^{s,bf} \right) - D_A^{s,af} \tag{3.16}$$

where l_{el} is the length of the element, a_{el}^f is the fatigue crack length within the element, the $D_A^{s,bf}$ is the static damage caused by the maximum fatigue load applied to the joint and the $D_A^{s,af}$ is the unwanted static damage. Based on Eq. 3.16 the rate of fatigue damage growth as a function of load cycle is defined as follows:

$$\frac{D_A^f}{dn} = \frac{1 - D_A^{s,bf}}{l_{el}} \frac{da}{dn} - \frac{D_A^{s,af}}{dn} \tag{3.17}$$

In another study, Harper and Hallett [12] proposed a fatigue degradation law based on the LDFA by considering the unwanted fatigue damage. They considered explicit finite element analysis using LS-Dyna to simulate the fatigue response of composite materials. Based on their model, unwanted fatigue damage is calculated as the difference between the integrated area under the traction separation curve with no fatigue damage and the area under the real traction separation law where the fatigue damage is considered. This unwanted fatigue damage is schematically shown in Fig. 3.2.

Accordingly, they defined the different damage mechanics and fracture mechanics parts of the LDFA as follows:

$$\frac{\partial d_f}{\partial dL_D} = \frac{1 - d_s - d_{fu}}{L_{el}} \tag{3.18}$$

$$\frac{\partial L_D}{\partial N} = \frac{L_{el}}{L_{fat}} \frac{\partial a}{\partial N} \tag{3.19}$$

However, in their model, the cohesive zone length (Lcz) should be extracted manually at the last static load increment.

By combining the two above equations and considering the LDFA relation, the following equation is obtained for the rate of damage propagation:

Fig. 3.2 Schematic of the fatigue damage and static damage considered by Harper and Hallett [12]

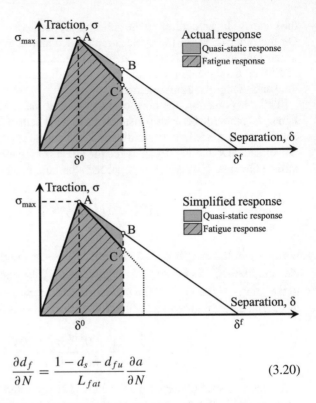

$$\frac{\partial d_f}{\partial N} = \frac{1 - d_s - d_{fu}}{L_{fat}} \frac{\partial a}{\partial N} \tag{3.20}$$

Recently Ebadi et al. [18], proposed a fatigue damage model considering the unwanted fatigue (static) damage. They employed a normalized strain energy Paris law model to simulate the crack propagation (Eq. 3.21).

$$\frac{\partial A}{\partial N} = \begin{cases} C\left(\frac{\Delta G}{G_c}\right)^m & G_{th} < G_{\max} < G_c \\ 0 & otherwise \end{cases} \tag{3.21}$$

The authors also included the static damage into the fatigue life predictions using Eq. 3.22 and as schematically shown in Fig. 3.3.

$$d_{N+\Delta N} = d_N + \Delta d_f + \Delta d_s \tag{3.22}$$

where d_N shows the damage in the previous time increment $d_{N+\Delta N}$ is the damage value in the current time increment. d_f is the fatigue damage and d_s is the unwanted fatigue (static) damage. Figure 3.3 schematically show the model used by Ebadi et al. [18].

Accordingly, they proposed the following relation for the rate of fatigue damage accumulation in a cohesive element:

Fig. 3.3 Schematic view of
the static and fatigue damage
parameters

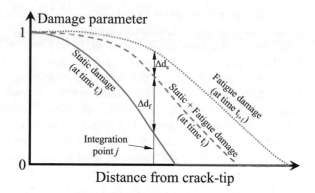

$$
\overline{\left(\frac{\partial d_f}{\partial N}\right)}_i = \frac{\left(\frac{\partial A}{\partial N}\right)_i}{(A_{cz})_i \left(1 + \frac{\overline{(\Delta d_s)_i}}{\overline{(\Delta d_f)_i}}\right)} \tag{3.23}
$$

where $\overline{(\Delta d_f)}$ and $\overline{(\Delta d_s)}_i$ are the average damage increments for the fatigue and static parts, respectively.

Based on Eq. 3.24, $\left(\frac{\partial d_f}{\partial N}\right)_i^J$ can be calculated approximately as below for each integration point:

$$
\left(\frac{\partial d_f}{\partial N}\right)_i^J \cong \frac{\left(\frac{\partial A}{\partial N}\right)_i}{(A_{cz})_i \left(1 + \chi_i^J\right)}; \quad for \; J \in Cohesive \; Zone \; Area \tag{3.24}
$$

where χ_i^J is obtained from the $(i-1)$th increment using the following relation:

$$
\chi_i^J = \frac{(\Delta d_s)_{i-1}^J}{(\Delta d_f)_{i-1}^J} = \frac{(\Delta d_s)_{i-1}^J}{\left(\frac{\partial d_f}{\partial N}\right)_{i-1}^J \times \Delta N} \quad i > 1 \tag{3.25}
$$

It was previously mentioned in this book that most of the LDFA models developed for fatigue life analysis of adhesives using CZM are a function of a specific length (area) called Acz (cohesive zone). It was also explained that this parameter can be manually or automatically obtained. However, the difficulties of measuring this zone, the key influence of the process zone size on the results and also the sensitivity of the length (area) of the cohesive process zone to the mesh size, has driven the development of models where the process zone length is not required in the LDFA models [20]. Based on the model of Kawashita and Hallet [20], the fatigue damage rate is in this case obtained by a non-local technique. Based on their approach, the delamination point should be first identified. Then, the local direction of damage propagation must be defined and, finally, the SERR for the failed element within the wake of the crack front needs to be calculated.

Another important point associated to LDFA models is to know which CZM parameters undergo degradation during the fatigue life. The above-mentioned models mostly apply the damage parameter to the stiffness of the cohesive element. For interface layers in which zero thickness cohesive elements are employed and where the initial stiffness is set to higher orders of magnitude, these approaches might lead to an inaccurate result. To overcome this issue, Pirondi and Moroni [21] recently proposed a model in which both the initial stiffness and the initial maximum traction were considered as variables. The authors believe that for high stiffness cohesive elements, consistent to those used in the interface cohesive elements, the definition of the damage based on the traction is more appropriate, especially from the physical point of view.

Apart from the LDFA models, some authors also developed and proposed empirical fatigue life prediction approaches based on the data fitting concepts. The data fitting method and the corresponding models are discussed in the next section.

3.2.2   Experimental Data Fitting (EXFIT)

Another load envelope technique suitable to simulate the fatigue behaviour of bonded joints is known as experimental data fitting (EXFIT). In EXFIT, the damage parameter is defined as a function of an effective stress [22] or strain [3, 5, 23] component. In terms of the numerical implementation, these approaches are generally simpler than the LDFA methods. However, EXFIT models are based on a trial-and-error procedure which is their main disadvantage. Different EXFIT models have been proposed in the literature to simulate the fatigue degradation properties of adhesive materials using a cohesive zone modelling approach.

The mode I fatigue crack growth of different adhesives was studied by Walander et al. [24], where a stress based model was defined to simulate the rate of fatigue damage evolution as a function of load cycles. As shown in Eq. 3.27, they used both the maximum stress and the stress threshold in their model.

$$\frac{\mathrm{d}D}{\mathrm{d}N} = \alpha \left(\frac{\frac{\sigma}{1-D} - \sigma_{th}}{\sigma_{th}} \right)^{\beta} \tag{3.26}$$

where s is the maximum stress at each cycle. s_{th}, α, and β are three fitting parameters that are used to fit the numerical results to the experimental data. Following the EXFIT concepts, Khoramishad et al. [23] proposed a strain approach to calculate the rate of damage as a function of load cycles. Similar to the work of Walander et al. [24], Khoramishad et al. [23] used three calibration parameters. Their model is based on the maximum strain and maximum threshold strain obtained at each loading cycle.

$$\frac{\Delta D}{\Delta N} = \begin{cases} C \times (\varepsilon_{\max} - \varepsilon_{th})^b & [if\ \varepsilon_{\max} \geq \varepsilon_{th}] \\ 0 & [if\ \varepsilon_{\max} \leq \varepsilon_{th}] \end{cases} \tag{3.27}$$

where, for mixed mode conditions, the maximum strain is obtained using the following relation:

$$\varepsilon_{max} = \frac{\varepsilon_n}{2} + \sqrt{\left(\frac{\varepsilon_n}{2}\right)^2 + \left(\frac{\varepsilon_s}{2}\right)^2} \tag{3.28}$$

To numerically implement their model, Khoramishad et al. [23] employed a bilinear shape CZM. Different approaches to numerically implement the fatigue models are discussed in the next chapter.

Later, Khoramishad et al. [25] extended their model by including two new parameters, g and n, used to take into account the effects of load ratio and maximum fatigue load in fatigue response of adhesive joints. Accordingly, their model was modified as follows:

$$\frac{\Delta D}{\Delta N} = \begin{cases} C \times \left[(\varepsilon_{max} - \varepsilon_{th}) \times \gamma^n \right]^b & [if \ \varepsilon_{max} \geq \varepsilon_{th}] \\ 0 & [if \ \varepsilon_{max} \leq \varepsilon_{th}] \end{cases} \tag{3.29}$$

where

$$\gamma = \frac{\frac{1-R}{2}}{1 - \left[\frac{P_{max}}{2P_s}(1+R) \right]^m} \cdot \left[\begin{matrix} m = 1 \ for \ the \ Goodman \ approach \\ m = 2 \ for \ the \ Gerber \ approach \end{matrix} \right] \tag{3.30}$$

In Eqs. 3.30 and 3.31, R is the load ratio, P_{max} is the maximum fatigue load, P_s is the static strength, and n is a calibration parameter. Hosseini-Toudeshky et al. [26] extended the work of Khoramishad et al. [23] by proposing a new relation including the mode mixity effects. Accordingly, drawing inspiration from the model of Blanco et al. [27] they calculated the fitting parameters as follows:

$$\log \alpha_m = \log \alpha_I + \beta_{mix} \log \alpha_{50\%} + \beta_{mix}^2 \log \frac{\alpha_{II}}{\alpha_{50\%}\alpha_I} \tag{3.31}$$

$$\beta_m = \beta_I + \beta_{50\%}\beta_{mix} + (\beta_{II} - \beta_I - \beta_{50\%})\beta_I^2 \tag{3.32}$$

$$\varepsilon_{thm} = \varepsilon_{thI} + (\varepsilon_{thII} - \varepsilon_{thI})(\beta_{mix})^\mu \tag{3.33}$$

where the subscript 50% denotes the mode ratio of 50%. β_{mix} is defined as follows:

$$\beta_{mix} = \frac{\sqrt{\gamma_{sn}^2 + \gamma_{tn}^2}}{\varepsilon_n} \quad for \ \varepsilon > 0 \tag{3.34}$$

One of the simplest forms of the EXFIT models was proposed by Johar et al. [28]. They suggested a power law based damage approach, defined as follows:

$$D_{cyc} = 1 - N^b \tag{3.35}$$

where b is a fitting parameter obtained experimentally. They applied the damage variable to all the three main cohesive parameters including the initial stiffness, maximum traction, and fracture energy in a triangular CZM shape. Most of the works discussed above use a simple bilinear (triangular) shape CZM. However, in the work of Pang et al. [29] a polynomial TSL was considered (See Eq. 3.36).

$$t_j = \frac{27}{4} T_j \left(\frac{u_j}{\delta_j} \right) \left(\frac{u_j}{\delta_j} - 1 \right)^2 . \quad \left[if \ u_j \le \delta_j . j = n.t.b \right] \tag{3.36}$$

Accordingly, they defined the fatigue damage model as follows:

$$\chi = \left(\frac{T^N}{T^0} \right)^{\frac{1}{3}} = (\beta N^\alpha + 1)^{\frac{1}{3}} \tag{3.37}$$

where j denotes the loading mode (I and II), T is the maximum traction, u is the current displacement and δ is the maximum separation at failure. χ is a dimensionless damage parameter and it is assumed to be the same for both pure modes I and II loading conditions. The parameters α and β are empirical coefficients that must be experimentally obtained. Based on Eq. 3.38, the damage parameter is defined as a ratio of the traction at load cycle N to the initial maximum traction of the cohesive elements. By considering Eqs. 3.37 and 3.38, they modified the polynomial TSL as follows:

$$t_j = \frac{27}{4} T_j \left(\frac{u_j}{\delta_j} \right) \left(\frac{u_j}{\delta_j} - \chi \right)^2 . \quad \left[if \frac{u_j}{\delta_j} \le \chi . \ j = n.t.b \right] \tag{3.38}$$

The authors degraded the fracture energy as a function of load cycle using the damage parameter defined in 3.38. According to their analysis, the value of fracture energy after N fatigue cycles can be calculated as follows:

$$G^N = G^0 \chi^4 \tag{3.39}$$

Robinson et al. [5] also considered the EXFIT approach to simulate the fatigue damage evolution at interfaces with the zero thickness of cohesive elements and for two-dimensional problems. They employed a strain-based function to calculate the accumulated fatigue damage.

$$\frac{\partial D_f}{\partial N} = \frac{C}{1 + \beta} e^{\lambda D} \left[\left(\frac{\delta_{max}}{\delta_a} \right)^{1+\beta} - \left(\frac{\delta_{min}}{\delta_a} \right)^{1+\beta} \right] \tag{3.40}$$

where β, c, and g are fitting parameters. δ_{max} and δ_{min} are the maximum value of the displacement component during the cycle and its corresponding minimum value, respectively. It should be noted that considering both the maximum and minimum displacements in Eq. 3.41 means that the effect of R ratio and maximum fatigue load are all taken into account in this model. However, the accuracy of the model for different R ratios and maximum fatigue load values was not analyzed by the authors.

A wear-out approach that can also be considered as an EXFIT model was proposed by Costa et al. [30] for pure-mode I loading conditions, where a bilinear CZM was used to simulate damage along the bonded line in a DCB joint. The authors suggested the following relationship to calculate the rate of degradation in maximum traction and fracture energy of cohesive elements as a function of the loading cycle:

$$y(N) = y_0 \left(1 - \frac{N}{N_f} \right)^k \tag{3.41}$$

In their proposed model, two constants (calibration parameters) were employed. The first one is k which is an exponent that controls the rate of reduction in cohesive properties, and the second one is N_f which considers the total fatigue life of the analyzed bonded joints. However, to obtain the N_f, the authors suggested the following relation:

$$N_f = \frac{\Delta a}{(da/dN)_a} \tag{3.42}$$

where $(da/dN)_a$ is an average fatigue crack growth rate, and Δa is the total bonded length.

By considering the constitutive equations for shear fatigue loading, the model was extended by Monteiro et al. [31] to pure mode II conditions, where a user element subroutine (See Sect. 4.4.2 for more information about the subroutines) was developed to implement the fatigue model in Abaqus. Using the same concepts, the model was later extended to mixed mode [32] but instead of the user element, a user material routine embedded in Abaqus was employed to perform the fatigue calculations. However, in the approach mentioned above, the N_f parameter must be known in advance, which is the main disadvantage of this fatigue life prediction model. To solve this problem, Akhavan-Safar et al. [33, 34] linked the considered EXFIT model with the LEFM using a Paris law relation. In their proposed model, N_f is automatically calculated using a user-developed material subroutine and considering the rate of crack growth calculated by Paris law. To achieve this, at each time increment and based on the current traction, damage, and separation status the current value of the SERR was calculated by considering the area under the CZM shape corresponding to the current values of traction and separation. Using the SERR value and knowing the fracture energy of the adhesive corresponding to its initial conditions, the da/dN rate can be calculated using the Paris law relation as follows:

$$\frac{da}{dN} = c_I \left(\frac{G_I}{G_{Ic}} \right)^{m_I} \tag{3.43}$$

where c_I and m_I are the Paris law fitting parameters for pure mode I loading conditions. G_I is mode I SERR.

However, it should be noted that the calibration parameters used in the Paris law must already be known and defined as input data. Consequently, N_f is automatically calculated for each material point and at each time increment. Using the N_f value, the maximum tensile and fracture energy is updated and subsequently, the degraded CZM is formed. Considering the updated form of the CZM and knowing the current separation, the damage value is calculated. Whenever the damage value reaches 1, or the updated value of the maximum traction or tensile fracture energy is equal to or less than zero, the corresponding point can be considered as failed. This approach was evaluated with pure mode I loading conditions using DCB [34] specimens. However, Akhavan-Safar et al. [33] later extended this approach to pure shear loading conditions, considering the constitutive laws of CZM for mode II.

3.3 Mixed Mode Paris Laws

TSLs are usually obtained for pure mode conditions but are used in practice to model mixed mode stress states. As discussed before, different fatigue models have been developed to estimate the fatigue life of adhesive joints in pure mode conditions using the LDFA concepts. This section discusses how the pure mode models should be merged to be appropriate for mixed mode fatigue loading. One of the major components of the LDFA is the LEFM based part where a Paris law relation is used. Although different Paris laws have been proposed, almost all of them use the two well-known fitting parameters called C and m. C and m are simply obtained for pure mode conditions using fatigue fracture tests, however, for mixed mode, fatigue crack growth data is necessary to know how the pure mode values should be mixed together to obtain mixed mode C and m values. One of the first mixing approaches was introduced by Blanco et al. [27] as follows:

$$\log_{10} C = \log_{10} C_I + \left(\frac{G_{II}}{G_T} \right) \log_{10} C_m + \left(\frac{G_{II}}{G_T} \right)^2 \log_{10} \left(\frac{C_{II}}{C_m C_I} \right) \tag{3.44}$$

$$m = m_I + m_m \left(\frac{G_{II}}{G_T} \right) + (m_{II} - m_I - m_m) \left(\frac{G_{II}}{G_T} \right)^2 \tag{3.45}$$

where I, II and m are representatives of mode I, mode II and mixed mode conditions, respectively, and the mode ratio (G_{II}/G_T) is defined based on the strain values as follows:

$$\frac{G_{II}}{G_T} = \frac{\beta^2}{1 + \beta^2} \tag{3.46}$$

where $\beta = \frac{\varepsilon_{sn}}{\varepsilon_n}$ is the ratio of shear strain to the normal strain.

This technique was considered by several authors such as [32] to simulate the fatigue life of bonded joints using CZM.

Kenane and Benzegaggh [35], using two material constant called m_d and m_b introduced their mixing method which is defined as follows:

$$d = d_I + (d_{II} - d_I) \left(\frac{G_{II}}{G_T} \right)^{m_d}$$

$$\ln B = \ln B_{II} + (\ln B_I - \ln B_{II}) \left(1 - \frac{G_{II}}{G_T} \right)^{m_b} \tag{3.47}$$

where d_I, B_I, d_{II}, B_{II} are the Paris law calibration parameters that should be obtained experimentally using pure mode I and pure mode II fatigue fracture tests. The values of m_d and m_b were respectively considered as 1.85 and 0.35 in their study.

Russel and Street [36] used the following relation to mix the pure mode results for mixed mode loading state:

$$c_m = c_I \left(\frac{G_I}{G_T} \right) + c_{II} \left(\frac{G_{II}}{G_T} \right) \tag{3.48a}$$

$$m_m = m_I \left(\frac{G_I}{G_T} \right) + m_{II} \left(\frac{G_{II}}{G_T} \right) \tag{3.48b}$$

In this relation, the subscript m denotes the calculated value while the subscripts I and II correspond to mode I and mode II results. This approach was considered by several authors such as [20]. Based on the BK method [35] and considering the 3D stress state condition, Oliveira and Donadon [19] included the mode III loading mode in the mixing role:

$$m = m_I + m_b \phi_{IIf} + m_c \phi_{IIIf} + m_d \phi_{IIf} \phi_{IIIf}$$
$$+ (m_{II} - m_I - m_b) (\phi_{IIf})^2 + (m_{III} - m_I - m_c) (\phi_{IIIf})^2$$
$$\log(c) = \log(c_I) + \log(c_b) \phi_{IIf} + \log(c_c) \phi_{IIIf}$$
$$+ \log(c_d) \phi_{IIf} \phi_{IIIf} + \log \left(\frac{c_{II}}{c_I c_b} \right) \phi_{IIf}^2 + \log \left(\frac{c_{III}}{c_I c_c} \right) \phi_{IIIf}^2 \tag{3.49}$$

where ϕ_{II} and ϕ_{III} are the mixed mode ratio for pure mode II and pure mode III loading conditions, defined as follows:

$$\phi_{II} = \frac{G_{II}}{G} \quad and \quad \phi_{III} = \frac{G_{III}}{G} \tag{3.50}$$

Mixed mode calibration parameters m_c and C_c are obtained from mixed mode I and III loading conditions and the parameters m_d and C_d must be extracted from mixed mode II/III experiments.

Not only the calibration parameters should be adapted for mixed mode conditions, but also the energy function used in the Paris law. One of the simplest mixing rules for this purposed was proposed by Russel and Street [36] where they simply defined the mixed SERR as the summation of the mode I and mode II parts of the SERR. Their model is mathematically described as follows:

$$G = G_I + G_{II} \tag{3.51}$$

Curley et al. [37] presented another simple form of the mixing role where they simply calculated the total SERR by Eq. 3.53.

$$\frac{dA}{dN} = B\Delta G_T^d = B(\Delta G_1 + \Delta G_2)^d \tag{3.52}$$

Another well-known method used for mixing the pure mode energy parameters was proposed by Benzeggagh and Kenane [35]. They suggested the following model for mixed mode I + II fatigue crack growth condition:

$$G_c = G_{Ic} + (G_{IIc} - G_{Ic})\left(\frac{G_{II}}{G_T}\right)^{\eta} \tag{3.53}$$

$$G_{th.mixed\ mode} = G_{Ith} + (G_{IIth} - G_{Ith})\left(\frac{G_{II}}{G_T}\right)^{\eta_2} \tag{3.54}$$

where subscripts I and II refer to mode I and mode II, respectively.

The mixed mode SERR was also calculated using the following equation [38, 39].

$$\Delta G_{eq} = \Delta G_I + \frac{\Delta G_{II}}{\Delta G_I + \Delta G_{II}}\Delta G_{II} \tag{3.55}$$

3.4 Variable Amplitude Fatigue Loading

Although most of the studies mentioned above address the evolution of fatigue damage to constant amplitude fatigue load conditions, this not the actual fatigue load that bonded structures generally experience in service. In service, the amplitude often changes with time. Variable amplitude loading can accelerate the rate of fatigue damage evolution or, in some cases, can cause fatigue crack retardation. Consequently, considering the effects of variable amplitude on the fatigue response of adhesive joints is of paramount importance. Variable amplitude loading is generally considered as a content amplitude block followed by an overload or, in some cases, different constant amplitude fatigue loading blocks with different load levels or R-ratios are considered.

For the cycle by cycle analysis discussed in the next section, each loading cycle is individually simulated, so that the effect of variable amplitude can be taken into account with less effort (further information can be found in [40, 41]). But for the cycle jumping or load envelope strategy in which only the maximum fatigue load is applied to the joint in a quasi-static condition, the damage accumulation models must be able to take the effects of variable amplitude into consideration and this might create important challenges to the analysis. Accordingly, in this section the load envelope based models, in which the effects of variable amplitude loading are included, are discussed.

The Palmgren–Miner model that is based on the summation of the fatigue damage for each loading block is one of the famous approaches used for variable amplitude loading. This method has been employed in some studies in combination with the LDFA to consider variable amplitude fatigue loading conditions [42]. A strength wear out approach was employed by Shenoy et al. [43] to consider the effects of variable amplitude. In this approach, the strength of the joint decreases with changes in the load amplitude. Of course, the amount of wear out in the strength depends on the influence of the load amplitude applied to the adhesive. The cycle mix technique is another variable amplitude fatigue life analysis method, considered by Erpolat et al. [44]. They reduced the residual strength as a function of mean fatigue load. The use of fracture mechanics based methods to include the fatigue crack growth under variable amplitude loading is another technique adopted in some studies [42, 45, 46]. However, it should be noted that the fracture mechanics based methods do not consider the initiation fatigue life, which may significantly underpredict the fatigue life.

Most of the studies on the fatigue life analysis of materials under variable amplitude loading are based on the continuum mechanics approach and a combination of variable amplitude fatigue loading with the CZM is less commonly encountered. One of these studies deals with the adaptation of an EXFIT model for variable amplitude fatigue loading conditions proposed by Khoramishad et al. [25]. The authors employed a bilinear TSL to simulate the damage evolution into the adhesive layer. To achieve this, they included both the maximum load and the R ratio into their model.

As discussed above, besides the load envelope strategy (EXFIT and LDFA models), the variable amplitude fatigue loading has also been considered for the cycle by cycle method [40, 41]. Two plane strain solid elements bonded with a layer of cohesive elements were simulated by Siegmund [40] to analyze the effect of variable amplitude loading using an irreversible CZM approach. According to their numerical results, for variable amplitude conditions, the accumulation of damage is significantly influenced by the sequence of the fatigue cycles. They found that moving from a low to a high amplitude or from a high to a low amplitude will lead to two different joint fatigue behaviors. Single overloads in a constant amplitude fatigue loading can also change the cyclic response of the bonded structures [41]. Although several methods have been proposed for fatigue life analysis of joints under variable amplitude loading, however, each model has several calibration parameters which require experimental data for their calibration. On the other hand, the models are still limited to specific loading, material, and geometry conditions. Accordingly, for a safe design against variable amplitude loading conditions, an appropriate safety factor should be applied to the obtained results.

3.5 Cycle-By-Cycle or Unloading–Reloading CZM

Although for high cycle fatigue regime where the load level is well below the yielding point of the joint, a load envelope strategy is a suitable choice to analyze the cohesive response of the fatigue loaded adhesives, but for higher load levels in which the fatigue life is short and loading history-dependent, a cycle-by-cycle analysis is recommended. A combination of the cycle-by-cycle strategy with the TSL has been studied by various authors using an irreversible CZM.

In cycle-by-cycle strategy, instead of using Paris law and loading envelope, the real unloading–reloading procedure is applied to the joint using an irreversible constitutive CZM law. The key point in this approach is that the unloading path in the CZM model is different from the reloading path [47, 48]. Using irreversible CZM dates back to the 1990s [49]. Considering the load-unload CZM, the dissipation strain energy was calculated using the area enclosed between the TSL curves. Figure 3.4 shows an initial irreversible CZM proposed by Horii et al. [50]. According to this model, the stiffness of the element is degraded with a degradation factor for the reloading section, as shown in Fig. 3.4. Using this method, the elastic shakedown phenomenon in which the stiffness of the cohesive element is the same for both the unloading and reloading path is prevented. By degrading the stiffness, the maximum tensile strength of the cohesive elements is also decreased. Preventing elastic shake

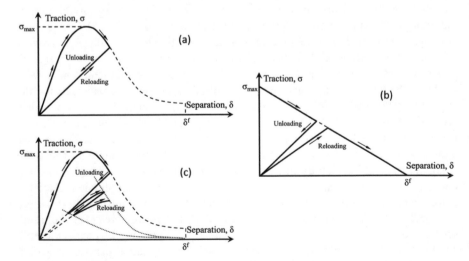

Fig. 3.4 Reversible (**a**) versus irreversible (**b**) and **c** CZM law

down makes it possible to measure the deformation energy, which is the energy enclosed between the reloading and unloading path. Several research works were conducted on the irreversible CZM for monotonic loading conditions, including the works of Needleman [51], Camacho and Ortiz [48, 52], and Siegmund [40].

Nguyen et al. [48] proposed one of the first irreversible TSL adapted for cycle by cycle fatigue simulation of cohesive elements.

Although an extensive study has been carried out on TSLs in a bilinear form, as discussed in Chap. 2, some authors believe that this type of CZM is limited to a specific material primarily with a linear behavior to failure. In terms of fatigue analysis using the irreversible technique (cycle by cycle), the use of other forms of TSLs would also be recommended in specific cases. One such form is the linear-exponential form which was already discussed in Chap. 2. This CZM form was considered by Serebrinsky and Ortiz [53] to simulate the hysteresis loops created during fatigue loading.

Most of the works conducted on the hysteresis dependent TSLs are mainly valid for pure mode I loading conditions, whereas, in practice, joints are designed to be loaded mainly in shear while the actual stress state follows a mixed mode. In this sense, it is necessary to evaluate the models developed for the shear loading modes. Roe and Siegmund [54] attempted to solve this problem by developing a model capable of simulating the load-unload behavior for different loading modes using an irreversible TSL. In his approach, the defined cohesive zone parameters can be obtained from the S–N results of the adhesive. To calibrate the model, the numerical results must be compared with the experimental data, something which is seen as the main disadvantage of its irreversible TSL method.

Using linear assumptions for the softening part of the CZM can simplify the model and make it easier to implement in FE codes. There is also an extensive study on the CZM with the linear damage evolution part that facilitates the achievement of the required constitutive laws. However, for very low cycle fatigue where the adhesive undergoes plastic deformation during fatigue, and specially for materials that show a non-linear response, the linear softening path may not be accurate enough. In this case, the softening path is defined in analogy to the plasticity behavior of the adhesive as discussed by Lemaitre [55]. The effects of 3D stress state, using XFEM and CZM for mixed mode, considering the stress constraint effects, and the gradient of plasticity, are other aspects considered in non-linear reloading in history dependent and irreversible CZMs [41, 56–59]. Considering mode I fatigue loading for adhesive materials and drawing inspiration from [47, 48, 54], Maiti and Geubelle [60] proposed a fatigue mode using a nonlinear reloading method. These authors employed a cohesive volumetric finite element approach to numerically implement their model, a numerical approach that makes it possible to solve large scale problems. The same authors later extended their work by considering the crack closure effects [60]. In their model the irreversible cohesive stiffness as a function of time is defined as follows:

Fig. 3.5 Scheme of the
Maiti and Geubelle [60]
irreversible CZM model

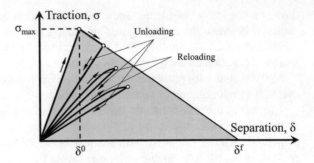

$$\dot{k}_{coh} = \begin{cases} -\frac{1}{\alpha} N_f^{-\beta} k_{coh} \dot{\Delta}_n & \left[if \ \dot{\Delta}_n \geq 0 \right] \\ 0 & \left[if \ \dot{\Delta}_n \leq 0 \right] \end{cases} \tag{3.56}$$

where N_f is the number of cycles that the material point experiences since the onset
of damage initiation where the normal stress reaches the maximum tensile traction.
α and β are material constants or we can call them as calibration factors that should
be experimentally obtained. $\dot{\Delta}_n$ is the rate of change in the normal separation. These
parameters as well as the unload-reload CZM developed by Maiti and Geubelle [60]
are schematically shown in Fig. 3.5.

The unloading path is often linear, however, it is important to know on which slope
the unloading occurs. One approach is to unload the element with a stiffness lower
than the initial stiffness in order to reach the zero stress/zero separation at a complete
unloading condition. Accordingly, the reduction in the slope in the unloading part
of irreversible CZMs will be different for different cycles. Another approach is to
unload the elements with a stiffness equal to the initial stiffness. Both approaches
are schematically shown in Fig. 3.6.

Another parameter commonly considered in the cycle-by-cycle strategy is the
friction level between the crack surfaces. It should be noted that for compressive
loads in the normal direction, contact between the crack surfaces will build up,
leading to dissipation of frictional energy and frictional stresses. Although most of
the work discussed above has neglected the effects of friction due to the complexity
of the problem, a simple form of contact friction has already been considered in a
limited set of studies on cycle-by-cycle TSLs, such as [47, 61].

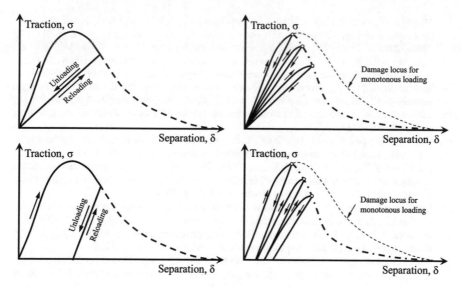

Fig. 3.6 Different unloading–reloading strategies

References

1. Turon, A., et al. 2007. Simulation of delamination in composites under high-cycle fatigue. *Composites Part A: Applied Science and Manufacturing* 38 (11): 2270–2282.
2. Van Paepegem, W., and J. Degrieck. 2001. Fatigue degradation modelling of plain woven glass/epoxy composites. *Composites Part A: Applied Science and Manufacturing* 32 (10): 1433–1441.
3. Peerlings, R.H., et al. 2000. Gradient-enhanced damage modelling of high-cycle fatigue. *International Journal for Numerical Methods in Engineering* 49 (12): 1547–1569.
4. Khoramishad, H., et al. 2010. Predicting fatigue damage in adhesively bonded joints using a cohesive zone model. *International Journal of Fatigue* 32 (7): 1146–1158.
5. Robinson, P., et al. 2005. Numerical simulation of fatigue-driven delamination using interface elements. *International Journal for Numerical Methods in Engineering* 63 (13): 1824–1848.
6. Rocha, A., et al. 2019. Paris law relations for an epoxy-based adhesive. *Proceedings of the Institution of Mechanical Engineers, Part L: Journal of Materials: Design and Applications* 1464420719886469.
7. Pirondi, A., and F. Moroni. 2010. A progressive damage model for the prediction of fatigue crack growth in bonded joints. *The Journal of Adhesion* 86 (5–6): 501–521.
8. Moroni, F., and A. Pirondi. 2011. A procedure for the simulation of fatigue crack growth in adhesively bonded joints based on the cohesive zone model and different mixed-mode propagation criteria. *Engineering Fracture Mechanics* 78 (8): 1808–1816.
9. De Moura, M., and J. Gonçalves. 2014. Development of a cohesive zone model for fatigue/fracture characterization of composite bonded joints under mode II loading. *International Journal of Adhesion and Adhesives* 54: 224–230.
10. De Moura, M., and J. Gonçalves. 2015. Cohesive zone model for high-cycle fatigue of composite bonded joints under mixed-mode I+ II loading. *Engineering Fracture Mechanics* 140: 31–42.
11. Pirondi, A., G. Giuliese, and F. Moroni. 2016. Fatigue debonding three-dimensional simulation with cohesive zone. *The Journal of Adhesion* 92 (7–9): 553–571.

12. Harper, P.W., and S.R. Hallett. 2010. A fatigue degradation law for cohesive interface elements–development and application to composite materials. *International Journal of Fatigue* 32 (11): 1774–1787.
13. May, M., and S.R. Hallett. 2011. An advanced model for initiation and propagation of damage under fatigue loading–part I: Model formulation. *Composite Structures* 93 (9): 2340–2349.
14. May, M., et al. 2011. An advanced model for initiation and propagation of damage under fatigue loading–part II: Matrix cracking validation cases. *Composite Structures* 93 (9): 2350–2357.
15. Rice, J.R. 1968. Mathematical analysis in the mechanics of fracture. *Fracture: An Advanced treatise* 2: 191–311.
16. Harper, P.W., and S.R. Hallett. 2008. Cohesive zone length in numerical simulations of composite delamination. *Engineering Fracture Mechanics* 75 (16): 4774–4792.
17. Naghipour, P., M. Bartsch, and H. Voggenreiter. 2011. Simulation and experimental validation of mixed mode delamination in multidirectional CF/PEEK laminates under fatigue loading. *International Journal of Solids and Structures* 48 (6): 1070–1081.
18. Ebadi-Rajoli, J., et al. 2020. Progressive damage modelling of composite materials subjected to mixed mode cyclic loading using cohesive zone model. *Mechanics of Materials* 103322.
19. de Oliveira, L.A., and M.V. Donadon. 2020. A cohesive zone model to predict fatigue-driven delamination in composites. *Engineering Fracture Mechanics* 235: 107124.
20. Kawashita, L.F., and S.R. Hallett. 2012. A crack tip tracking algorithm for cohesive interface element analysis of fatigue delamination propagation in composite materials. *International Journal of Solids and Structures* 49 (21): 2898–2913.
21. Pirondi, A., and F. Moroni. 2019. Improvement of a cohesive zone model for fatigue delamination rate simulation. *Materials* 12 (1): 181.
22. Van Paepegem, W., J. Degrieck, and P. De Baets. 2001. Finite element approach for modelling fatigue damage in fibre-reinforced composite materials. *Composites Part B: Engineering* 32 (7): 575–588.
23. Khoramishad, H., et al. 2010. A generalised damage model for constant amplitude fatigue loading of adhesively bonded joints. *International Journal of Adhesion and Adhesives* 30 (6): 513–521.
24. Walander, T., et al. 2014. Fatigue damage of adhesive layers–experiments and models. *Procedia Materials Science* 3: 829–834.
25. Khoramishad, H., et al. 2011. Fatigue damage modelling of adhesively bonded joints under variable amplitude loading using a cohesive zone model. *Engineering Fracture Mechanics* 78 (18): 3212–3225.
26. Hosseini-Toudeshky, H., et al. 2020. Prediction of interlaminar fatigue damages in adhesively bonded joints using mixed-mode strain based cohesive zone modeling. *Theoretical and Applied Fracture Mechanics* 106: 102480.
27. Blanco, N., et al. 2004. Mixed-mode delamination growth in carbon–fibre composite laminates under cyclic loading. *International Journal of Solids and Structures* 41 (15): 4219–4235.
28. Johar, M., M.S.E. Kosnan, and M.N. Tamin. 2014. Cyclic cohesive zone model for simulation of fatigue failure process in adhesive joints. In *Applied Mechanics and Materials*. Trans Tech Publ.
29. Pang, J., et al. 2013. Fatigue analysis of adhesive joints under vibration loading. *The Journal of Adhesion* 89 (12): 899–920.
30. Costa, M., et al. 2018. A cohesive zone element for mode I modelling of adhesives degraded by humidity and fatigue. *International Journal of Fatigue* 112: 173–182.
31. Monteiro, J., et al. 2019. Mode II modeling of adhesive materials degraded by fatigue loading using cohesive zone elements. *Theoretical and Applied Fracture Mechanics* 103: 102253.
32. Rocha, A., et al. 2020. Numerical analysis of mixed-mode fatigue crack growth of adhesive joints using CZM. *Theoretical and Applied Fracture Mechanics* 102493.
33. Akhavan-Safar, A., et al. 2021. A modified degradation technique for fatigue life assessment of adhesive materials subjected to cyclic shear loads. *Proceedings of the Institution of Mechanical Engineers, Part C: Journal of Mechanical Engineering Science* 235 (3): 550–559.

34. Akhavan-Safar, A., et al. 2020. Tensile fatigue life prediction of adhesively bonded structures based on CZM technique and a modified degradation approach. *Proceedings of the Institution of Mechanical Engineers, Part G: Journal of Aerospace Engineering* 234 (13): 1988–1999.
35. Benzeggagh, M.L., and M. Kenane. 1996. Measurement of mixed-mode delamination fracture toughness of unidirectional glass/epoxy composites with mixed-mode bending apparatus. *Composites Science and Technology* 56 (4): 439–449.
36. Russell, A.J., and K.N. Street. 1989. Predicting interlaminar fatigue crack growth rates in compressively loaded laminates. In *Composite Materials: Fatigue and Fracture, Second Volume*. ASTM International.
37. Curley, A., et al. 2000. Predicting the service-life of adhesively-bonded joints. *International Journal of Fracture* 103 (1): 41–69.
38. Quaresimin, M., and M. Ricotta. 2006. Life prediction of bonded joints in composite materials. *International Journal of Fatigue* 28 (10): 1166–1176.
39. Quaresimin, M., and M. Ricotta. 2006. Stress intensity factors and strain energy release rates in single lap bonded joints in composite materials. *Composites Science and Technology* 66 (5): 647–656.
40. Siegmund, T. 2004. A numerical study of transient fatigue crack growth by use of an irreversible cohesive zone model. *International Journal of Fatigue* 26 (9): 929–939.
41. Jiang, H., X. Gao, and T.S. Srivatsan. 2009. Predicting the influence of overload and loading mode on fatigue crack growth: A numerical approach using irreversible cohesive elements. *Finite Elements in Analysis and Design* 45 (10): 675–685.
42. Shenoy, V., et al. 2010. Fracture mechanics and damage mechanics based fatigue lifetime prediction of adhesively bonded joints subjected to variable amplitude fatigue. *Engineering Fracture Mechanics* 77 (7): 1073–1090.
43. Shenoy, V., et al. 2009. An evaluation of strength wearout models for the lifetime prediction of adhesive joints subjected to variable amplitude fatigue. *International Journal of Adhesion and Adhesives* 29 (6): 639–649.
44. Erpolat, S., et al. 2004. A study of adhesively bonded joints subjected to constant and variable amplitude fatigue. *International Journal of Fatigue* 26 (11): 1189–1196.
45. Ashcroft, I. 2004. A simple model to predict crack growth in bonded joints and laminates under variable-amplitude fatigue. *The Journal of Strain Analysis for Engineering Design* 39 (6): 707–716.
46. Erpolat, S., et al. 2004. Fatigue crack growth acceleration due to intermittent overstressing in adhesively bonded CFRP joints. *Composites Part A: Applied Science and Manufacturing* 35 (10): 1175–1183.
47. Yang, B., S. Mall, and K. Ravi-Chandar. 2001. A cohesive zone model for fatigue crack growth in quasibrittle materials. *International Journal of Solids and Structures* 38 (22–23): 3927–3944.
48. Nguyen, O., et al. 2001. A cohesive model of fatigue crack growth. *International Journal of Fracture* 110 (4): 351–369.
49. Kuna, M., and S. Roth. 2015. General remarks on cyclic cohesive zone models. *International Journal of Fracture* 196 (1–2): 147–167.
50. Horii, H., H.C. Shin, and T.M. Pallewatta. 1992. Mechanism of fatigue crack growth in concrete. *Cement and Concrete Composites* 14 (2): 83–89.
51. Needleman, A. 1992. Micromechanical modelling of interfacial decohesion. *Ultramicroscopy* 40 (3): 203–214.
52. Camacho, G.T., and M. Ortiz. 1996. Computational modelling of impact damage in brittle materials. *International Journal of Solids and Structures* 33 (20–22): 2899–2938.
53. Serebrinsky, S., and M. Ortiz. 2005. A hysteretic cohesive-law model of fatigue-crack nucleation. *Scripta Materialia* 53 (10): 1193–1196.
54. Roe, K., and T. Siegmund. 2003. An irreversible cohesive zone model for interface fatigue crack growth simulation. *Engineering Fracture Mechanics* 70 (2): 209–232.
55. Lemaitre, J. 2012. *A Course on Damage Mechanics*. Springer Science & Business Media.
56. Wang, B., and T. Siegmund. 2005. A numerical analysis of constraint effects in fatigue crack growth by use of an irreversible cohesive zone model. *International Journal of Fracture* 132 (2): 175–196.

57. Brinckmann, S., and T. Siegmund. 2008. Computations of fatigue crack growth with strain gradient plasticity and an irreversible cohesive zone model. *Engineering Fracture Mechanics* 75 (8): 2276–2294.
58. Xu, Y., and H. Yuan. 2009. Computational modeling of mixed-mode fatigue crack growth using extended finite element methods. *International Journal of Fracture* 159 (2): 151–165.
59. Liu, J., H. Yuan, and R. Liao. 2010. Prediction of fatigue crack growth and residual stress relaxations in shot-peened material. *Materials Science and Engineering: A* 527 (21–22): 5962–5968.
60. Maiti, S., and P.H. Geubelle. 2006. Cohesive modeling of fatigue crack retardation in polymers: Crack closure effect. *Engineering Fracture Mechanics* 73 (1): 22–41.
61. Yang, B., and K. Ravi-Chandar. 1998. A single-domain dual-boundary-element formulation incorporating a cohesive zone model for elastostatic cracks. *International Journal of Fracture* 93 (1): 115–144.

Chapter 4
Numerical Simulation

Abstract By knowing the shape and fatigue model of the CZM discussed in Chaps. 2 and 3, it is possible to start simulating the fatigue of bonded joints. But before that, we need to be familiar with the different aspects of a numerical CZM-based fatigue life analysis and this is the main goal of this chapter. Due to their popularity, many commercial programs have already included cohesive elements and cohesive laws in their material and element libraries. However, there are some differences between the CZMs used in these FE codes, as discussed in this chapter. According to the procedure discussed above, the first step in this process is to choose the CZM shape, defining the cohesive properties, and assigning these properties to the adhesive layer. At this stage, users need to be familiar with the stress state in cohesive elements, damage propagation path, mesh strategy used with cohesive elements, etc. Two different numerical strategies can be followed to define the cohesive layer. The most common approach is to define a finite cohesive layer thickness with cohesive elements. The second strategy is to use cohesive contact. All these aspects are discussed in this chapter. CZM fatigue models are not usually listed in the material library of FE programs and users therefore need to embed the fatigue analysis into the FE codes using subroutines. Different subroutines are available in Abaqus and can be employed for this purpose. USDFLD, UMAT, and UEL are the most common of these. All these routines are discussed in this chapter. Finally, the chapter is concluded by presenting some examples where bonded steel substrates were analyzed in terms of fatigue using a load envelope fatigue model and by the aim of UMAT and UEL subroutines.

4.1 Introduction

In Chap. 1, different types of fatigue loads and various factors that affect the fatigue response of bonded structures were discussed. Such information is necessary in order to select a suitable CZM shape for the fatigue analysis. It is also crucial to possess information about the CZM shape, its specifications, and its constitutive laws. All of these points were covered in Chap. 2. Chapter 2 also includes an explanation on how a suitable CZM should be selected. However, to numerically simulate the fatigue response of the adhesive, one key part is still missing, which is the fatigue model itself, where the variation of the cohesive properties as a function of the loading cycle is defined. This part was discussed in Chap. 3, where different CZM fatigue methodologies were reviewed. Now, with this information, one can start the process of simulating adhesive joints under cyclic loading conditions. But the numerical part itself also has important and key aspects that should be discussed. The goal of the current chapter is to discuss these key aspects of the numerical simulation of bonded joints under cyclic loading conditions and by using cohesive elements. In the first section, the different commercial software packages used for cohesive analysis are briefly introduced. However, since Abaqus is widely used as an FE package for CZM analysis of bonded structures, the use of this FE software and its specificities are further discussed in this chapter.

4.2   Commercial Software

Due to the extensive application of the CZM in damage analysis of structures, commercial FE codes such as Abaqus and ANSYS have already included cohesive elements in their libraries. Apart from these two well-known FE programs, there are also alternative FE codes such as Warp3D and Zebulon in which we can find some cohesive based models for damage analysis of structures. However, all these programs are still quite limited in terms of the type of cohesive elements and cohesive laws they support. Accordingly, to apply the recent TSLs reviewed in Chaps. 2 and 3, users need to develop some user-defined routines to be able to include the customized CZMs . In terms of the shape of the TSL, Abaqus supports bilinear, exponential, and trapezoidal laws. It also covers the mixed-mode conditions. The same is true for Ansys, although some differences can be found in the models available in the two programs). COMSOL, a commercial multiphysics software, has also included cohesive elements in its library. However, the elements present in COMSOL are mainly used for delamination and interfacial failure analysis.

The consideration of mixed-mode conditions is another key feature in FE codes suitable for cohesive modeling of adhesive joints. In some programs, such as Franc2D only the 2D condition can be simulated and in some others such as the WARP3D just the 3D problems can be defined.

In the following section the most key features in numerical simulation of fatigue in bonded joints using CZM are discussed.

4.3   CZM

To perform a numerical simulation of a joint subjected to fatigue loads and using CZM, two main steps are considered. The first step is to choose an appropriate CZM shape. This is generally similar to the approach used for a static simulation and therefore, the same procedure must be followed. The second major step is to modify the CZM or the analysis to take into account the effects of cyclic loadings.

Consequently, the general numerical CZM settings are first discussed in this section and the details on fatigue simulation of adhesive joints using CZM are presented in the next section (Sect. 4.4).

4.3.1 Parameters Adjustments

As stated in the introduction of this chapter, the parameters discussed in this section are mainly focused on the Abaqus FE software, although they are not totally exclusive to it.

- Element type

For cohesive analysis, the cohesive layer must be defined using cohesive elements. Solid elements under 3D conditions have nine stress (strain) components (three normal and six shear components), while for cohesive elements under 3D conditions only three stress components (one normal and two shear components) are defined. This is one of the main differences between cohesive elements and regular solid elements. Figure 4.1 shows the difference between stress states in a solid element and a cohesive element under 3D conditions.

Consequently, the linear assumption for the initial part of the CZM the tensile separation relationship in cohesive elements is calculated as follows. According to this relationship, the stiffness matrix (K) relates the tension with the separation for each direction. In the case of coupled stress and shear, the relationship is defined as follows:

$$t = \left\{ \begin{array}{c} t_n \\ t_s \\ t_t \end{array} \right\} = \left[\begin{array}{ccc} K_{nn} & K_{ns} & K_{nt} \\ K_{sn} & K_{ss} & K_{st} \\ K_{tn} & K_{ts} & K_{tt} \end{array} \right] \left\{ \begin{array}{c} \delta_n \\ \delta_s \\ \delta_t \end{array} \right\} = K\delta \qquad (4.1)$$

However, for the decoupled conditions where there is no interaction between the normal stiffness and the shear components, the constitutive law above will be

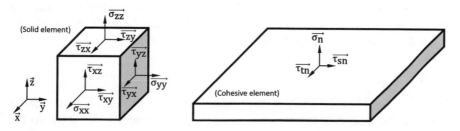

Fig. 4.1 Solid element versus cohesive element

simplified as follows, where only the initial stiffness for pure modes must be defined.

$$
t = \left\{ \begin{array}{c} t_n \\ t_s \\ t_t \end{array} \right\} = \left[\begin{array}{ccc} K_{nn} & & \\ & K_{ss} & \\ & & K_{tt} \end{array} \right] \left\{ \begin{array}{c} \delta_n \\ \delta_s \\ \delta_t \end{array} \right\} = \boldsymbol{K}\boldsymbol{\delta} \tag{4.2}
$$

For a cohesive element analysis, where a layer of cohesive elements with a specific thickness is used along the bond line, the initial stiffness for the tensile and shear direction can be considered as, respectively, the ratio of Young's modulus and shear modulus, both divided by the thickness of the adhesive. However, for the interface cohesive element or for cohesive contact where the cohesive thickness might be considered as zero, the initial stiffness as a penalty parameter should be much higher than the tensile stiffness of the adhesive.

To mesh the cohesive layer, a swept meshing technique is used where the mesh is defined on one side of the specimen and then the nodes of the elements on the source face are copied along the path defined by the user. For the CZM, the direction defined should be along the thickness of the cohesive elements to create a correct orientation for the defined cohesive layer.

• Damage propagation path

To mesh the cohesive layer, the sweep mesh technique is used, where the mesh is defined on one side of the sample and then the element nodes on the source face are copied along the user-defined path. For CZM, the defined direction must be along the thickness of the cohesive elements, allowing to create a correct orientation for the defined cohesive layer.

• Energy versus displacement damage evolution

Another important aspect that must be defined in the numerical analysis of bonded joints is the damage evolution method. Two options are available in Abaqus but in other FE programs, only one of these methods might be available. One is to use displacement, where the damage is defined as a function of the element's equivalent displacement. The energy approach is another technique, significantly more common. In this method fracture energies for pure modes are set to form the CZM shape and to calculate the damage evolution within the cohesive layer. In this case, the CZM is formed by Abaqus to ensure that the area under the shape of the CZM is equal to the fracture energies defined by the user. Using a tabular option, available in Abaqus, it is also possible to customize the damage evolution. The tabular damage evolution technique is also used for mixed-mode conditions where custom CZM shapes are used.

• CZM shape

Due to its simplicity, the study by Hillerborg et al. [1] is widely used in interfacial analysis. However, it is not accurate for many cases, especially for more ductile conditions and when a finite thickness is considered for cohesive elements. Consequently, the bilinear, trapezoidal, and the exponential TSLs that are found in almost all FE software packages, representing the most employed CZM shapes.

Although almost all CZM shapes often have convergence problems, a problem that is more prominent for the triangle shape compared to the exponential TSL. Passing the maximum tensile point causes a significant change in the slope of the stiffness of the cohesive elements, a change which is smoother in the exponential or trapezoidal CZM. However, in some of the trapezoidal shapes and in exponential CZM, the cohesive element stiffness reaches zero in some regions which, under certain loading conditions, can also lead to numerical problems. Convergence issues are discussed in the following sections.

• Mode mixity

The way that pure mode results are mixed to form a CZM shape for different mode ratios is a key part in CZM analysis of bonded joints. Different mixed mode models were introduced in Chap. 2. However, FE codes are limited in terms of the mixed mode approaches available. In Abaqus, to mix the two pure modes I and II an effective displacement is defined as follows:

$$\Delta_m = \sqrt{\Delta_n^2 + \Delta_s^2} \tag{4.3}$$

where Δ_n and Δ_t are the normal and shear parts of the mixed mode displacement. To measure the effective separation, the strain values along the normal and shear direction reported by Abaqus can be employed. The traction values are also shown directly as the stress components reported in the FE code results. For the complete failure, using the BK method the following relation is used by Abaqus:

$$\delta_{mf} = \frac{2}{K\delta_c}\left[G_n + (G_t - G_n)\left(\frac{\beta e^2}{1 + \beta e^2}\right)^{\eta}\right] \tag{4.4}$$

where η is a material constant that should be obtained experimentally. When both the normal and shear fracture energies are the same, the total mixed mode fracture energy using the BK method is simplified as follows:

$$\delta_{mf} = 2\frac{G_n}{K\delta_c} \tag{4.5}$$

The power law, BK, and tabular format are the three options available in Abaqus for mixed mode damage evolution. The most finely detailed approach among these three methods is the tabular method in which the mixed mode CZM can be defined manually in a tabular format based on the real response of the joint, although some manual calculation might be needed. In contrast, using the BK and the Power law method, Abaqus automatically calculates the damage evolutions with minimal effort.

- Convergence problems

Facing convergence problems when analyzing bonded joints using CZM is common. In this sense, some techniques are used aiming at the convergence of the analysis. One of the tools that must be considered by users to solve this issue is a precise control of the time increment. Users can define the step time and the time increment at each step of the analysis. The time increment in the analysis should be small enough to avoid the divergence problem at the point where damage starts within the bondline. Another approach to avoid convergence problems is to improve the shape of the CZM. In the triangle shape TSL, element stiffness changes significantly at the onset of damage, while in the exponential or cubic form of CZM the change in stiffness is much smoother. Another technique to avoid divergence problems is to define the viscosity or a damping factor. The viscous stabilizer makes the tangent stiffness matrix of the cohesive element positive for a small increment during the softening procedure. This viscous time increment is too small to introduce considerable errors into the results and the small viscosity helps to converge the analysis during the softening section. Consequently, the tensile decreases and increases repeatedly while the separation continuously increases along the softening part of the CZM form. Increasing the number of iterative and cutbacks in an increment of time can also help the analysis overcome some convergence issues. As a rule of thumb, it is best to avoid using fixed time increments and let the FE program adjust the time increments automatically. However, it should be noted that for the fatigue analysis using the cycle jumping strategy in which a static step is defined to simulate the fatigue behavior, the time increment represents the load cycles. Under these conditions, the fixed time increment can also be considered if no convergence issue occurs.

4.3.2 Cohesive Contact Versus Cohesive Element

There are two approaches to define the cohesive behavior of adhesives in a bonded joint. The most common approach is to use cohesive elements. In the cohesive elements method, cohesive elements with a finite thickness are used between the two substrate interfaces. Another technique for defining cohesive behavior is to establish a cohesive interaction or cohesive contact. The cohesive interface is defined through the interaction module in Abaqus where stiffness and cohesive damage characteristics are defined. Figure 4.2 schematically shows the two approaches. In Abaqus, both methods are available, while some FE programs are limited to just one of these techniques. Generally, both approaches use similar concepts and constitutive cohesive laws. For example, their linear behavior in the early part of the TSL is similar and they also employ similar damage initiation and evolution rules. In terms of simulation, using cohesive contact is easier than the cohesive element approach. On the other hand, in the cohesive contact method, the behavior of the cohesive part can be properly simulated using a wide range of interactions.

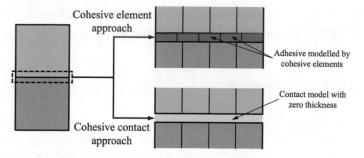

Fig. 4.2 The differences between cohesive element and cohesive contact

However, it should be noted that since the cohesive layer thickness is considered to be zero in this simulation, it is more suitable for modeling very thin bonded layers (compared to other joint dimensions) such as interfacial failure in composite materials. It should also be noted that using cohesive contact, the effect of adhesive thickness on the results is ignored. If the thickness of the adhesive layer is not negligible compared to other joint dimensions, the use of cohesive elements (not cohesive contacts) is recommended. One of the main advantages of the cohesive contact compared to the cohesive element is that the cohesive contact can be used for surfaces that are not initially bonded. However, once they are linked during the analysis, the cohesive contact will analyze the damage evolution through the new linked zone. On the other hand, if the already failed surfaces come into contact again during the analysis, there is the possibility in this approach to do a post-fail analysis, since regular contact is also defined for the cohesive contact surfaces. Since no cohesive element is used in the cohesive contact approach, different calculation points are considered in the two approaches. For cohesive elements, the cohesive integration points are the points where calculations are made. In the case of cohesive contact, the calculations are made on the contact constraints that are associated with the slave nodes defined on the contact surfaces. On contact surfaces, the finer meshed surface generally acts as a slave surface. Therefore, in terms of mesh refinement, the finer the mesh of a surface, the greater the resolution of the spatial variables. The mesh refinement is different in the cohesive element approach, in that the user can easily refine the mesh size by defining more nodal seeds along with the connecting line regardless of the number of substrate nodes in the interfaces.

Another point that must be considered is the interaction of the regular contact defined in the FE software and the cohesive response of this contact. For normal stresses, no interaction between the regular contact and the cohesive contact is observed. On the other hand, in purely compressive loads, only regular contacts are active. However, for shear loads, the interaction of cohesive contact and regular contact is different. In a shear mode, for an undamaged condition, the constitutive laws of cohesive contact define the behavior of the joint. However, for damaged conditions, the cohesive elements of defined friction and shear interact. In this case, even after a complete rupture of the cohesive part, the friction defined in the regular contact will govern the behavior of the contact surface.

4.3.3 Mesh Size Sensitivity

As discussed above, depending on the type of analysis (cohesive contact vs. cohesive element), different techniques must be employed to refine the mesh size along with the cohesive layer. However, it should be noted that compared to other types of elements such as solid elements, the stress state within the adhesive layer is less sensitive to mesh size using cohesive elements. However, even so, the size of the elements along the bondline must be tightly controlled and kept small enough to obtain a more accurate result. There are some rules for the size (number) of the mesh along the adhesive layer. As cohesive elements are designed primarily for interfaces, the thickness of the cohesive element must be thin enough. As discussed in Chap. 3, the size of the cohesive zone (A_{cz} for 3D and L_{cz} for 2D) is employed in LDFA approaches for analyzing the fatigue life of adhesive joints. The size of the cohesive elements must be small enough to include enough elements within the process zone. As a general rule, the size of the cohesive elements should be small enough compared to the characteristic length of the samples. The characteristic length here can be the size of the plastic zone in front of the crack tip for the material considered. Consequently, for materials with a small-scale yield ahead of the crack tip, a smaller element size should be used compared to more ductile adhesives. It is also recommended not to use elements with a thickness greater than their length.

4.4 Fatigue Analysis

As discussed in Sect. 4.1, In terms of the numerical simulation, the fatigue life analysis of bonded joints using CZM approaches has two major steps. The first step is to choose and define a TSL which is discussed in Sect. 4.3, and the second step is to apply a suitable fatigue model to the selected TSL. Accordingly, in this section, the most important aspects for the second step are reviewed.

4.4.1 Cycle by Cycle Versus Cycle Jumping

As discussed in Chap. 3, two strategies can be followed to implement incremental fatigue degradation of cohesive elements. The strategy most often considered by researchers is the cycle jump method (loading envelope). In this approach, a specific number of load cycles must be defined for each time increment in the developed routine. For the second method, the analysis is carried out cycle by cycle. In the loading strategy, two static steps are defined. In the first step, the load is increased from zero to the maximum fatigue load. In the second step, the stress must be kept constant and equal to the maximum fatigue load. The time in the first increment is not meaningful, but for the second step, the time indicates the cycle number. Consequently, the time in the second step should be sufficient to make sure it can cover the full fatigue life of the joint. In this regard, a cycle jump parameter must be defined. The cycle jump parameter determines the number of cycles that correspond to each time increment in the second analysis step. The expected total fatigue life of the joint must be less than the number of time increments times the cycle jumping factor. The mathematical expression of this statement is as follows:

$$Total\ fatigue\ life < cycle\ jumping\ factor \times number\ of\ increments \tag{4.6}$$

In the load envelope method, the maximum load is kept constant. Thus, at each time increment, the maximum energy value corresponding to the updated properties of the cohesive elements and based on the current values of the equivalent separation is calculated. However, this value is the maximum SERR for the current load cycle. For the LDFA and also the EXFIT methods, the Paris law can be a function of both the minimum SERR and its maximum values. Thus, the minimum SERR values must also be calculated at each cycle. The minimum SERR corresponds to the minimum load, while in the analysis only the maximum load is applied. To solve this problem, the following relationship is employed, where the minimum SERR is related to the maximum SERR using the load ratio (R ratio).

$$(G_{max} - G_{min}) = G_{max}(1 - R^2)\ or\ \frac{G_{min}}{G_{max}} = R^2 \tag{4.7}$$

Some authors [2] believe that this relation is not a precise assumption for adhesive materials but, it can significantly simplify the simulation procedure.

Monteiro et al. [2] showed that the relation between the minimum and maximum SERR does not necessarily follow Eq. 4.7. According to their results, the obtained values of G_{min}/G_{max} are different from R^2 for the tested adhesives. As mentioned by Monteiro et al. [2], a possible reason is the crack closure phenomenon. The authors found that the difference between G_{min}/G_{max} and R^2 increases by increasing the load level at similar R ratios. On the other hand, it was shown that the difference between the G_{min}/G_{max} and R^2 decreases by keeping the load level constant and at the same time increasing the R ratio. The difficulties associated to the experimental measurement of G_{min} during the unloading cycles is another reason that leads to the fact that the R ratio in bonded joints might not be related to the ratio of the G_{min}/G_{max} as postulated in Eq. 4.7.

For very low cycle fatigue, considering a cycle-by-cycle analysis can provide a more accurate estimate of fatigue life, since the same loads encountered in real practical applications are applied to the model. However, due to its simplicity, a sinusoidal cyclic load is usually considered in the literature. Thus, the number of cycles applied must be almost known a priori. Except for the loading conditions, the remaining steps including the implementation of the constitutive laws in the FE program are similar to the loading envelope strategy. Similar to the load envelope strategy, for the cycle-by-cycle technique, the constitutive CZM laws must be embedded in the FE programs using auxiliary codes. Normally, in Abaqus, UMAT and UEL can be employed and are discussed in the next section. Some authors have considered UEL to analyze the life of bonded joints based on CZM and cycle-by-cycle methodology. Using an irreversible CZM, Roe and Siegmund [3] simulated fatigue crack growth along an interface. They used a four-node element in a simulation where two integration points for each element were considered. In their analysis, the damage variable was averaged per element. Using a custom element, they were able to implement the irreversible CZM model developed in Abaqus.

4.4.2 User Subroutines

Although FE codes have a list of available CZM models to choose from for static analyses, almost no CZM-based fatigue models are included in these programs. Thus, users must develop their own fatigue models and then implement them in the FE software. There are several possible approaches to implement a CZM based fatigue model. Commercial FE programs such as Abaqus offer users this possibility to create their own custom material response, where the fatigue behavior of adhesives in a bonded structure can be defined. To this end, in Abaqus users can develop different subroutines. Subroutines are generally Fortran codes, employed by Abaqus as part of the FE analysis process. Using a Fortran code, the value of the field is calculated and reported to the Abaqus and accordingly, the cohesive properties corresponding to this field value are used to analyze the damage within the cohesive elements. USDFLD (User Defined Field) is one of the techniques used by some authors to analyze the CZM-based fatigue behavior of adhesives. In a USDFLD, a relationship is made between a defined field variable and cohesive properties. A USDFLD allows you to define variables in material points and as a function of time. However, the USDFLD routine has access to material points only at the beginning of each time increment, which means that it must follow an explicit approach. This means that the properties and status of the element in each increment are the initial values at the beginning of that increment and are not influenced by analysis during that increment. Consequently, time increment settings play a key role in this type of analysis. Still, attracted by its simplicity, several authors have used this routine to successfully implement CZM fatigue models in FE programs. Pirondi et al. [4] used the USDFLD routine to modify the cohesive properties by setting a field variable. They combined this subroutine with URDFIL which is used for the post-processing of the results during analysis. Using the time increments and number of cycles defined for each increment, they calculated the number of cycles corresponding to each damage status. They used Eq. 4.7 to calculate the minimum SERR for each cycle. The calculated

SERR in each cycle was compared with the strain limit energy. The analysis starts running if the calculated SERR is greater than the threshold value.

Hosseini-Toudeshky et al. [5] also considered the USDFLD routine to implement the CZM fatigue model in Abaqus, using a triangular CZM shape. Based on the numerical procedure, the equivalent separation value was calculated by the developed subroutine considering the strain values reported by Abaqus. Using this data and based on the fatigue model, the damage evolution rate was calculated. The results were then transferred to Abaqus in order to calculate the stress components and to obtain the rate of change in stress values as a function of the rate of change in the strain components. Giuliese et al. [6] incorporated the USDFLD routine into Abaqus to simulate a 3D fatigue response of bonded areas based on the LDFA method. The authors applied the damage value calculated on the initial stiffness of the cohesive elements. To do so, they considered a cycle jump method where only the maximum fatigue load is applied to the model as static load. To calculate the modal ratio, an energy-based approach was employed (to find more information on the differences between energy-based and separation techniques in modal-ratio analysis see Sect. 2.4.2). USDFLD has also been employed by other authors [7, 8] especially where the property wear approach for fatigue degradation of cohesive elements is used. Except for the USDFLD routine, there are other types of routines in Abaqus that have been considered for fatigue analysis of bonded areas based on the CZM method. One of them that is an advanced subroutine is the User Defined Mechanical Material Behavior (UMAT). UMAT is used to define material constitutive laws. UMAT calculations are applied to all material integration points. Using a UMAT, the user must define the Jacobian matrix of the model named DDSDDE. The Jacobian matrix is defined as follows:

$$DDSDDE(i.j) = \frac{\partial \Delta \sigma_i}{\partial \Delta \varepsilon_j} \tag{4.8}$$

where $\Delta\sigma$ is the true stress increment and $\Delta\varepsilon$ denotes the strain increments. DDSDDE defines the variation in the ith component of stress as a function of change in the jth strain component.

Due to its flexibility in defining the material behaviour, the UMAT has been considered in many studies to estimate the fatigue life of bonded joints. For example, Rocha et al. [9] employed UMAT to simulate the cyclic response of bonded joints. They extended the Costa et al. method [10] for mixed mode loading conditions. Using the cycle jumping technique and by considering a wear out approach they calculated the rate of degradation in the cohesive properties. The wear out approach in their analysis was applied to both the maximum traction and the fracture energy and the same rate of reduction for both the mode I and mode II cohesive properties was assumed. Using the updated values of the cohesive properties the stress values were updated using the DDSDDE matrix and the new stress values were passed to the Abaqus for further analysis in the following time increment.

UMAT was also employed by Akhavan-Safar et al. [11, 12] to predict the total fatigue life in DCB and ENF specimens subjected to pure mode I and pure mode II sinusoidal loads, respectively. By improving a previously published fatigue model, the authors successfully linked the wear approach and LEFM concepts using a Paris law model to automate the life prediction procedure.

User element (UEL) is one of the most powerful yet most complex routines available in Abaqus, having also been considered in some fatigue studies [10, 13]. Using UEL makes it possible to work at the element level. In the aforementioned routines (USDFLD and UMAT) the cohesive element is already defined in Abaqus CAE and the properties are defined using the routines, while in UEL not only the cohesive properties but also the cohesive elements must be defined using the subroutine. So, in terms of numerical procedure, the UEL is more complicated and needs more calculations and, consequently, more user coding in UEL is needed. Using an UEL, the behavior of the elements must be fully defined using the stiffness matrix. Consequently, Eq. 4.9 must be solved using the FE solver:

$$[K] \times \{d\} = \{f\} \tag{4.9}$$

where $[K]$ is the stiffness matrix, and $\{d\}$ and $\{f\}$ are the displacement vector and the vector of the external forces, respectively. In the UEL subroutine, each of the stiffness matrix and the external force must be obtained separately using the following relationships:

$$[K] = w[B]^T [T_d][B] \tag{4.10}$$

$$\{f\} = w[B]^T \{T\} \tag{4.11}$$

where w is the element width and $[B]$ is the global displacement-separation matrix. $\{T\}$ and $[T_d]$ are vector and matrix, respectively, indicated based on the CZM shape and traction–separation laws. Several shapes of the CZM discussed in Chaps. 2 and 3 can be considered to form the $\{T\}$ and $[T_d]$. In case of triangle and for pure mode I and II loading conditions, the following relations for the $\{T\}$ and $[T_d]$ were employed by Costa et al. [10] and Monteiro et al. [13], respectively.

Pure mode I

$$\{T\} = \left\{ \begin{array}{c} 0 \\ t(d)_I \end{array} \right\} \quad [T_d] = \left[\begin{array}{cc} 0 & 0 \\ 0 & t'(d)_I \end{array} \right] \tag{4.12}$$

Pure mode II

$$\{T\} = \left\{ \begin{array}{c} t(d)_{II} \\ 0 \end{array} \right\}; \quad [T_d] = \left[\begin{array}{cc} t'(d)_{II} & 0 \\ 0 & 0 \end{array} \right] \tag{4.13}$$

Fig. 4.3 Cohesive zone element in global (x, y) and local (ξ, η) coordinates

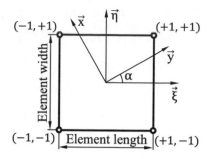

where t(d) corresponds to the equation that defines the triangular shape CZM and the indices I and II correspond to mode I and mode II, respectively. In the case of mixed-mode I/II, a combination of both Eqs. 4.12 and 4.13 must be considered.

In their simulation, Monteiro et al. [13] used a 4-node linear element in UEL to simulate the fatigue response in an ENF joint. Consequently, four shape functions must be defined. However, the number of nodes in each element can be increased to 6 or 8, which obviously increases computational time. The authors used a cohesive element of zero thickness in their analysis and, consequently, only the shape functions for nodes 1,4 and 2,3 must be obtained (see Fig. 4.3).

Considering the coordinate of ξ shown in Fig. 4.3, the shape functions can be defined as follows:

$$N_{1.4} = \frac{1}{2}(1 - \xi); \quad N_{2.3} = \frac{1}{2}(1 + \xi) \tag{4.14}$$

where [N] is the shape function matrix and is determined as follows:

$$[N] = \begin{bmatrix} N_{1.4} & & N_{2.3} & & N_{2.3} & & N_{1.4} & \\ & N_{1.4} & & N_{2.3} & & N_{2.3} & & N_{1.4} \end{bmatrix} \tag{4.15}$$

Another matrix that should be defined at the element level in UEL is the strain displacement that is shown by [B]:

$$[B] = [R][N] \tag{4.16}$$

where [R] is the transformations matrix obtained by considering the global to local coordinates and is defined as below:

$$[R] = \begin{bmatrix} \cos\alpha & \sin\alpha \\ -\sin\alpha & \cos\alpha \end{bmatrix} \tag{4.17}$$

where α is the angle between the two mentioned coordinate systems (see Fig. 4.3).

The remaining matrices are the {T} and [T_d] that, as mentioned above, are functions of the considered CZM shape. To define the {T} and [T_d], t(d) should be defined

Fig. 4.4 Different zones in a triangle (bilinear) CZM

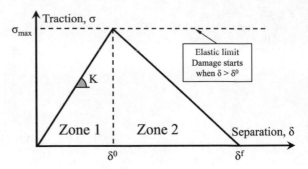

first. For a triangular TSL, $t(d)$ and t(d) are defined the same for both mode I and mode II as follows:

$$t_1(d) = \frac{t_m d}{d_0} \tag{4.18}$$

$$t_1'(d) = \frac{t_m}{d_0} \tag{4.19}$$

$$t_2(d) = t_m \left(1 - \frac{d - d_0}{d_f - d_0} \right) \tag{4.20}$$

$$t_2'(d) = \frac{-t_m}{d_f - d_0} \tag{4.21}$$

where the subscripts 1 and 2 denote the zones 1 and 2, respectively, as shown in Fig. 4.4. t_m is the maximum traction.

4.5 Examples of Numerical Models

In this section, examples are provided where the fatigue life of adhesively bonded steel substrates are analyzed using CZM and a property wear-out based fatigue model. A cyclic sinusoidal load was applied to all the simulated joints. An epoxy adhesive with the properties given in Table 4.1 was considered in this analysis.

Table 4.1 Mechanical properties of the adhesive and the substrate

Properties	Substrate	Adhesive
Maximum tensile strength (MPa)	–	31.3
Maximum tensile strain (%)	–	10.4
Young's modulus (MPa)	210,000	1159
Poisson's ratio (v)	0.3	

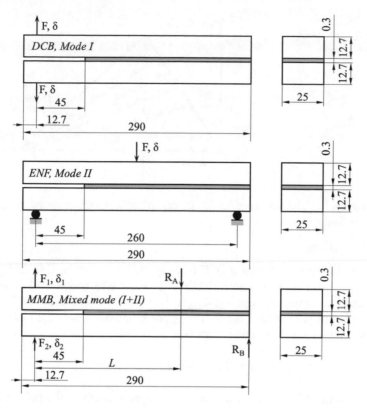

Fig. 4.5 Geometry of the analyzed joints (dimensions in mm)

Pure mode I, pure mode II, and mixed mode loading were analyzed. DCB samples with the same size were used for all the mentioned mode mixities (see Fig. 4.5). As shown in Fig. 4.5, a pre crack with the size of 45 mm was considered in the bondline. Two different fatigue models were employed including a wear-out approach and a combined wear-out and LEFM technique. These fatigue models were recently developed and evaluated by Costa et al. [10] and Monteiro et al. [13] for the wear out approach and [11, 12] for the linked wear-out/LEFM method. A mode I triangular shaped CZM was used to simulate the damage behaviour within the bonded area for all the considered conditions. Degradation was applied to both maximum tensile and fracture energy, as shown schematically in Fig. 4.6. A UMAT subroutine was developed and implemented in Abaqus for this study.

A cycle jumping strategy was followed, where two static steps are defined to simulate the fatigue loading conditions. In the first analysis step, the load increases until it reaches the maximum fatigue load. In the second step of the analysis, the load is held constant as the maximum fatigue load. The time increment is not meaningful in the first step, but in the second step, the time increment is used to calculate the

Fig. 4.6 Triangular
traction–separation law

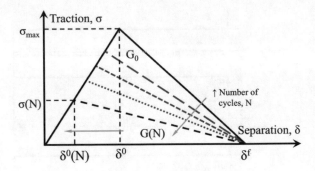

number of cycles. The details of the numerical analysis and the results for each mode
mix are as follows:

- Mode I loading

For mode I loading conditions, a DCB (see Fig. 4.5) was simulated in Abaqus. The
values of the maximum traction and mode I fracture energy was set to 31.3 MPa and
2.2 N/mm, respectively. The applied fatigue load was set to 60% of the static strength
of the joint. For pure mode I the following constitutive law was used to relate the
cohesive traction to the tensile separation.

$$\sigma_I = K_0[\delta_I - d(\delta_I + \langle -\delta_I \rangle)] \tag{4.22}$$

where $\langle \ \rangle$ is the McCauley bracket, σ_I is the tensile stress in the point and δ_I is
the corresponding strain. K_0 is the initial adhesive's stiffness and d is the damage
parameter.

The following relation, already discussed in Chap. 3, was used to simulate the
rate of fatigue degradation of the cohesive properties. The degradation was applied
solely to the maximum traction and the fracture energy.

$$y(N) = y_0 \left(1 - \frac{N}{N_f} \right)^k \tag{4.23}$$

$$N_f = \frac{\Delta a}{(da/dN)} \tag{4.24}$$

where Δa is the total length of the bonded area and (da/dN) is automatically calculated by UAMT and using Paris law relation (Eq. 4.25) at each time increment and
for the integration point where the energy value is higher than the threshold energy.

$$\frac{da}{dN} = c_I \left(\frac{G_I}{G_{Ic}} \right)^{m_I} \tag{4.25}$$

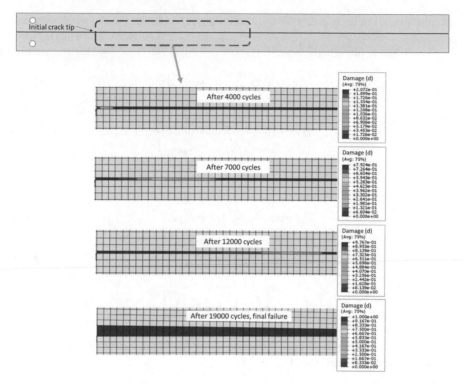

Fig. 4.7 Fatigue damage evolution in a DCB specimen loaded in pure tensile conditions using a linked wear out/LEFM model

where c_I and m_I are the Paris law fitting parameters. Degradation was applied to both the traction and fracture energy. Figure 4.7 shows the numerical results at different loading cycles. See [12] for more details.

- Mode II loading

The procedure described for mode I was also followed for mode II. However, in shear loading, the negative shear stresses are also able to create damage as do the positive shear stresses. Consequently, the following constitutive law was employed:

$$\tau = K_0[\delta_{II} - d\delta_{II}] \tag{4.26}$$

In terms of the loading conditions, the same strategy was also used (cycle skip), however, instead of an opening load, a three-point bending load was applied to the joint. Figure 4.5 schematically shows the loading conditions. There are two different shear strains (in-plane and out-of-plane shear components) in a 3D analysis. To calculate the shear stress usually an equivalent shear separation (δ_{shear}) is defined, based on the in plane, δ_{II}, and the out of plane, δ_{III}, shear displacement components (See Eq. 4.29). However, in a 2D analysis the out of plane shear strain of the cohesive elements is set to zero. Accordingly, it is only needed to set the out of plane shear strain (separation) to zero in the following equations for 2D conditions. In this section, a 2D problem is presented.

$$\tau = K_0[\delta_{shear} - d\delta_{shear}] \tag{4.27}$$

$$d = \frac{\delta_{shear}^f \left(|\delta_{shear}| - \delta_{shear}^0\right)}{|\delta_{shear}| \left(\delta_{shear}^f - \delta_{shear}^0\right)} \tag{4.28}$$

where:

$$|\delta_{shear}| = \sqrt{\delta_{II}^2 + \delta_{III}^2} \tag{4.29}$$

$$\delta_{shear}^0 = \left[\left(\delta_{II}^0\right)^2 + \left(\delta_{III}^0\right)^2\right]^{1/2} \tag{4.30}$$

$$\delta_{shear}^f = \left[\left(\delta_{II}^f\right)^2 + \left(\delta_{III}^f\right)^2\right]^{1/2} \tag{4.31}$$

Figure 4.8 shows the damage evolution along the bondline for different loading cycles. For more details, see [11].

• Mixed mode

For mixed-mode loading conditions, a wear out technique was followed, where the cohesive properties of the elements are decreased as a function of the load cycles (See Eq. 4.32). This function has already been proposed by Costa et al. [10] According to this approach, the N_f (total fatigue life of the specimen) as input data must be known a priori. Similar to mode I and mode II, the same joint was simulated, but under mixed-mode loading conditions (45°) (see Fig. 4.5).

$$y(N) = y_0 \left(1 - \frac{N}{N_f}\right)^k \tag{4.32}$$

where

$$N_f = \frac{\Delta a}{(da/dN)_a} \tag{4.33}$$

where the $(da/dN)_a$ is the average fatigue crack growth obtained experimentally using fatigue fracture tests, and the Δa is the total bonded area.

The following relation was defined between the traction and separation in the developed UMAT:

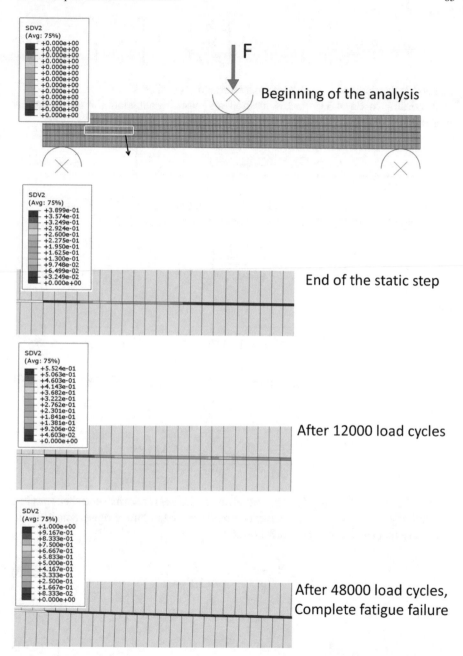

Fig. 4.8 Fatigue damage evolution in an ENF specimen loaded in pure mode II using a linked wear out/LEFM model (SDV2 here is the damage parameter)

$$\sigma = \begin{bmatrix} \sigma_I \\ \sigma_{II} \\ \sigma_{III} \end{bmatrix} = (1-d)K_0 \begin{bmatrix} \delta_I \\ \delta_{II} \\ \delta_{III} \end{bmatrix} - dK_0 \begin{bmatrix} \langle -\delta_I \rangle \\ 0 \\ 0 \end{bmatrix} \tag{4.34}$$

where d is the damage value and K is the adhesive's stiffness.

According to Eq. 4.35, the BK method was used to calculate the total fracture energy.

$$G_c = G_{Ic} + (G_{shearc} - G_{Ic}) \left[\frac{G_{shear}}{G_T} \right]^{\mu} \tag{4.35}$$

where $G_{shear} = G_{II} + G_{III}$, $G_T = G_I + G_{shear}$, and $G_{shearc} = \left(G_{IIc}^2 + G_{IIIc}^2 \right)^{0.5}$.

Damage for mixed mode bilinear CZM was calculated as follows:

$$d = \frac{\delta^f \left(\|\delta_{aeq}\| - \delta^0 \right)}{\|\delta_{aeq}\| \left(\delta^f - \delta^0 \right)} \tag{4.36}$$

where:

$$\|\delta_{aeq}\| = \sqrt{(\delta_I)^2 + (\delta_{shear})^2} \quad and \quad \delta_{shear} = \sqrt{\delta_{II}^2 + \delta_{III}^2} \tag{4.37}$$

$$\delta^{0^2} = (\delta_I)^2 + (\delta_{II})^2 + (\delta_{III})^2 = \delta_I^{0^2} + \left[\delta_{shear}^0{}^2 - \delta_I^{0^2} \right] \chi^{\eta} \tag{4.38}$$

, and

$$\delta^f = \frac{\delta_I^0 \delta_I^f + \left(\delta_{shear}^0 \delta_{shear}^f - \delta_I^0 \delta_I^f \right) \chi^{\eta}}{\delta^0} \tag{4.39}$$

where $\delta_I{}^0$ and $\delta_{shear}{}^0$ are the damage initiation conditions for normal and shear modes, respectively, and δ_I^f and δ_{shear}^f are displacements at final failure. χ represents the ratio G_{shear}/G_T that can be obtained as follows:

$$\chi = \frac{G_{shear}}{G_T} = \frac{\beta^2}{1 + 2\beta^2 - 2\beta} \tag{4.40}$$

where:

$$\beta = \frac{\delta_{shear}}{\delta_{shear} + \langle \delta_I \rangle} \tag{4.41}$$

Considering all the above-mentioned relations and the mixed mode fatigue life of the bonded specimen was calculated. Figure 4.9 shows the results. More details are provided in reference [9].

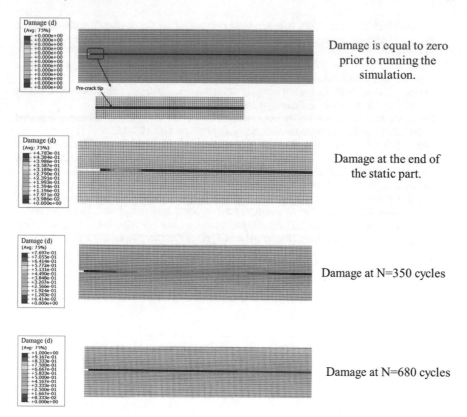

Fig. 4.9   Damage evolution in a bonded joint subjected to mixed mode loading using CZM and a wear out fatigue model

References

1. Hillerborg, A., M. Modéer, and P.-E. Petersson. 1976. Analysis of crack formation and crack growth in concrete by means of fracture mechanics and finite elements. *Cement and Concrete Research* 6 (6): 773–781.
2. Monteiro, J., et al. 2020. Influence of mode mixity and loading conditions on the fatigue crack growth behaviour of an epoxy adhesive. *Fatigue & Fracture of Engineering Materials & Structures* 43 (2): 308–316.
3. Roe, K., and T. Siegmund. 2003. An irreversible cohesive zone model for interface fatigue crack growth simulation. *Engineering Fracture Mechanics* 70 (2): 209–232.
4. Pirondi, A., G. Giuliese, and F. Moroni. 2016. Fatigue debonding three-dimensional simulation with cohesive zone. *The Journal of Adhesion* 92 (7–9): 553–571.
5. Hosseini-Toudeshky, H., et al. 2020. Prediction of interlaminar fatigue damages in adhesively bonded joints using mixed-mode strain based cohesive zone modeling. *Theoretical and Applied Fracture Mechanics* 106: 102480.
6. Giuliese, G., A. Pirondi, and F. Moroni. 2014. A cohesive zone model for three-dimensional fatigue debonding/delamination. *Procedia Materials Science* 3: 1473–1478.

7. Khoramishad, H., et al. 2010. A generalised damage model for constant amplitude fatigue loading of adhesively bonded joints. *International Journal of Adhesion and Adhesives* 30 (6): 513–521.
8. Khoramishad, H., et al. 2010. Predicting fatigue damage in adhesively bonded joints using a cohesive zone model. *International Journal of Fatigue* 32 (7): 1146–1158.
9. Rocha, A., et al. 2020. Numerical analysis of mixed-mode fatigue crack growth of adhesive joints using CZM. *Theoretical and Applied Fracture Mechanics* 102493.
10. Costa, M., et al. 2018. A cohesive zone element for mode I modelling of adhesives degraded by humidity and fatigue. *International Journal of Fatigue* 112: 173–182.
11. Akhavan-Safar, A., et al. 2021. A modified degradation technique for fatigue life assessment of adhesive materials subjected to cyclic shear loads. *Proceedings of the Institution of Mechanical Engineers, Part C: Journal of Mechanical Engineering Science* 235 (3): 550–559.
12. Akhavan-Safar, A., et al. 2020. Tensile fatigue life prediction of adhesively bonded structures based on CZM technique and a modified degradation approach. *Proceedings of the Institution of Mechanical Engineers, Part G: Journal of Aerospace Engineering* 234 (13): 1988–1999.
13. Monteiro, J., et al. 2019. Mode II modeling of adhesive materials degraded by fatigue loading using cohesive zone elements. *Theoretical and Applied Fracture Mechanics* 103: 102253.

Chapter 5
Summary and Conclusions

Abstract In Chaps. 1–4, the fatigue response of adhesive joints under cyclic loading was discussed in great detail, eventually explaining how a cohesive zone modelling (CZM) can precisely simulate the response of bonded structures. Different CZM shapes were introduced and their constitutive laws, their advantages, and their possible limitations were pointed out. It was also discussed that using CZM concepts, the fatigue response of bonded joints can be simulated. Different strategies in fatigue life analysis of adhesive joints using CZM based models were reviewed in Chap. 4. In this chapter a summary of all these discussions besides a conclusion are presented. Considering the points mentioned in Chaps. 1–4, it was shown that it is possible to accurately estimate the fatigue life of bonded joints using CZM based approaches if the problems are accurately defined using precise material properties and by choosing appropriate fatigue model and settings in FE programs.

5.1 Summary

Since the last decades of the twentieth century, synthetic adhesives have grown in their capabilities finding use in an increasingly larger range of applications, from everyday use to load-bearing structural components in highly technologically advanced products such as aircraft and automobiles. The better fatigue performance attained in comparison with other traditional joining methods (as also recently reported by Antelo et al. [1]) is one of the main reasons for the wide application of adhesives in the transportation sector. However, the fatigue response of bonded structures is still a very challenging topic and not yet fully explored.

Based on reported data, 90% of reported failures in industries are caused by fatigue damage and fatigue failure almost leads to a sudden component fracture. Consequently, considering the fatigue response of adhesive joints is an essential part of a joint design process. The total fatigue life is divided into two main stages, namely the initiation of fatigue life and fatigue crack propagation. Although in most applications, as a conservative approach, the crack initiation life is considered as the total fatigue life, however, to allow for lighter structures and for cost reduction, the crack growth fatigue life of adhesive joints should also be analyzed and considered in joint design in some advanced applications. The fatigue in bonded structures can be analyzed experimentally, numerically, or using a combination of these two approaches. Experimental methods can be divided into two groups, including fatigue characterization using standard or routine testing and fatigue life analysis of real components. Many factors are known to affect joint response to fatigue, including the loading parameters, joint geometry, and environmental conditions. All of these aspects were discussed in Chap. 1.

As experimental fatigue tests are very expensive and time-consuming, as discussed in Chap. 1, numerical techniques were employed to predict the fatigue behavior of bonded joints. Several techniques based on continuum mechanics, fracture mechanics, and damage mechanics have been proposed and used to estimate the fatigue response of bonded structures. Methods based on continuum mechanics mainly focus on the initiation fatigue life while approaches based on the fracture mechanics concepts consider the second stage of fatigue life which is fatigue crack growth. Damage mechanics concepts can be adapted to model both the fatigue life initiation and the fatigue crack propagation. Accordingly, damage mechanics based approaches have recently found greater adaption in research activities, with several damage models being developed and implemented in FE software packages. One of these approaches is CZM. CZM was first proposed in the 1960s and since then, has undergone many modifications and improvements. According to the CZM, traction in a cohesive zone is a function of separation and a damage parameter. The first constitutive law for the cohesive models was proposed by Hillerborg [2] but since then many different shapes of CZM such as trapezoidal, exponential, polynomial, etc. have been proposed. Considering the variety of the developed TSLs, a question that arises here is which CZM is suitable for a given joint considering the materials properties, loading conditions, joint geometry, etc. Chap. 2 discussed all these points and reviewed the properties and capabilities of the most common TSLs mentioned above. However, another question quickly rises, which is how different parameters in the considered TSL can be identified.

The different techniques available to shape a TSL were reviewed in Chap. 2, including the classical methods of characterization where fracture tests and strength experiments are conducted to obtain the cohesive properties of an adhesive, an inverse technique where the cohesive properties are set to fit the numerical results with the experimental data, and finally the direct approach in which using specific experiments the real shape of the TSL can be obtained. The last approach is thought to be the most precise one, since for the first two approaches the shape of the CZM should be selected before the test. In Chap. 2, the constitutive laws of different CZMs were also presented and discussed. Using the constitutive laws, cohesive models can be implemented in FE programs. Although the selection of an appropriate CZM and the identification of its cohesive parameters are important parts of the fatigue life analysis of bonded joints, it is still necessary to obtain further information to be able to simulate the fatigue life of the joints. This brings us to a point where one only possesses a CZM model that can simulate the quasi-static response of the joint and to analyze the fatigue behaviour one must (and define) know how cohesive properties vary (degrade) with load cycles. Chapter 3 addressed these points where different CZM based fatigue life prediction models were reviewed.

As discussed in Chap. 3, cohesive based fatigue methods can be classified into two main groups including the loading envelope (cycle jumping) strategy and the second one is the cycle-by-cycle analysis approach. The first one is mainly used for high cycle fatigue regimes and the second one is more suitable for low cycle fatigue where the fatigue stress level is close or above the yielding point of the adhesive.

According to this model, the fatigue loading can be simulated by a static load. To achieve this, two static steps are defined where in the first one the load is increased until reaching the maximum fatigue load and then it is kept constant in the second step. Two techniques called LDFA and EXFIT follow the load envelope strategy. In LDFA a link is created between the damage mechanics and fracture mechanics. Accordingly, the LDFA based models consist of two linked parts. Several models based on the LDFA have been developed by proposing new damage mechanics or by modifying the fracture mechanics part. Another load envelope technique that was discussed in Chap. 3 is EXFIT where the fatigue damage parameter is defined as a function of an effective stress or strain parameter. Accordingly, the EXFIT models are based on a trial-and-error procedure which makes them less efficient compared to the LDFA. In terms of the numerical implementation, the models based on the LDFA need further work than the EXFIT approaches. In addition to these techniques, a cycle-by-cycle method was also considered in some studies. Simulating actual fatigue cycles rather than using a constant static load is the main difference between load envelope strategies and the cycle-by-cycle approach. Since all cycles must be simulated in this technique, therefore, it is not an appropriate method for a high cycle fatigue regime. Cycle-by-cycle fatigue analysis is based on the concepts of irreversible CZMs, where cohesive stiffness in reloading is not the same as in unloading. Several authors have developed cycle-by-cycle CZMs. All of these techniques and their relationships were discussed in Chap. 3.

Now, by knowing the shape and fatigue model of the CZM, it's time to start simulating the fatigue of bonded joints. But before that, one must be familiar with the different aspects of a numerical CZM-based fatigue life analysis. This step was covered in the final Chap. 4. Due to their wide range of applications, many commercial programs have already included cohesive elements and cohesive laws in their material and element libraries. However, there are some differences between the CZMs used in these FE codes, as discussed in Chap. 4. Among these programs, Abaqus is a well-known FE software package, widely used in research activities in the literature, and thus was more deeply discussed in Chap. 4.

According to the procedure discussed above, the first step is to choose the CZM shape, define the cohesive properties, and assign these properties to the adhesive layer. At this stage, users must be familiar with the stress state in cohesive elements, damage propagation paths, mesh strategy used with cohesive elements, etc. All these aspects were discussed in Chap. 4. Two different numerical strategies can be followed to define the cohesive layer. The most common approach is to define a finite cohesive layer thickness with cohesive elements. The second strategy is to use cohesive contact. Although similar concepts of damage initiation and damage propagation and constitutive laws are used in both techniques, there are some differences between the two approaches. For example, cohesive contact is best suited with very thin layers, such as interfaces in composite materials. Using the cohesive contact method, the effect of adhesive thickness on the results is not considered. Cohesive contact can also be used for surfaces that are not initially bonded.

No CZM fatigue models are currently listed in the material library of FE programs. Therefore, users need to embed the fatigue analysis into the FE codes using subroutines. Different subroutines in Abaqus are available that can be employed for this purpose, with USDFLD, UMAT, and UEL being the most commonly used subroutines. Finally, Chap. 4 concludes with the presentation of some examples where bonded steel substrates were analyzed in terms of fatigue using a load envelope fatigue model and UMAT and UEL subroutines. The considered joints were subjected to pure mode I, pure mode II, and mixed-mode cyclic loading conditions. Joint geometry and dimensions, material properties, and constitutive laws were also provided in this section. In these examples, a triangle shape TSL was considered.

5.2 Conclusions

It is possible to accurately estimate the fatigue life of bonded joints using the CZM approach if the problem is accurately defined using precise material properties. However, it should be noted that without a careful definition of the problem, it is not possible to arrive at a good prediction of fatigue life. The CZM analysis of bonded structures, particularly for joints subjected to cyclic loading, has several facets that must be considered and mastered by a joint designer. Choosing an appropriate CZM form, using a suitable fatigue model to take into account the effects of loading conditions, and also considering the specificities of FE programs, are the main steps in a cohesive fatigue simulation. Although several attempts have been made to propose an accurate and more general cohesive based fatigue prediction model for bonded structures, fatigue life analysis of adhesive joints still suffers from the lack of a comprehensive model. Various parameters such as loading conditions (frequency, R-ratio, mean stress, etc.), environmental parameters (temperature and humidity), joint geometry (joint type and joint dimensions), and substrate materials can significantly influence the fatigue life of bonded joints and change the cohesive properties of the adhesives. Thus, in the analysis steps mentioned above, all these parameters must be taken into account.

References

1. Antelo, J., et al. 2021. Fatigue life evaluation of adhesive joints in a real structural component. *International Journal of Fatigue* 153: 106504.
2. Hillerborg, A., M. Modéer, and P.-E. Petersson. 1976. Analysis of crack formation and crack growth in concrete by means of fracture mechanics and finite elements. *Cement and Concrete Research* 6 (6): 773–781.

Printed in the United States
by Baker & Taylor Publisher Services